GIRLS, STYLE, AND SCHOOL IDENTITIES

GIRLS, STYLE, AND SCHOOL IDENTITIES

DRESSING THE PART

Shauna Pomerantz

palgrave
macmillan

First published in 2008 by
PALGRAVE MACMILLAN™
175 Fifth Avenue, New York, N.Y. 10010 and
Houndmills, Basingstoke, Hampshire, England RG21 6XS
Companies and representatives throughout the world.

PALGRAVE MACMILLAN is the global academic imprint of the Palgrave
Macmillan division of St. Martin's Press, LLC and of Palgrave Macmillan
Ltd. Macmillan® is a registered trademark in the United States, United
Kingdom and other countries. Palgrave is a registered trademark in the
European Union and other countries.

ISBN-13: 978-1-4039-8206-3
ISBN-10: 1-4039-8206-6

Library of Congress Cataloging-in-Publication Data is available from the
Library of Congress.

Pomerantz, Shauna.
Girls, style, and school identities : dressing the part / Shauna Pomerantz.
    p. cm.
ISBN 1-4039-8206-6 (alk. paper)

  1. Schoolgirls—British Columbia—Vancouver—Social conditions—
Case studies.   2. Schoolgirls—Clothing—British Columbia—
Vancouver—Case studies.   3. Girls—Education (Secondary)—British
Columbia—Vancouver—Case studies.   4. Women—Identity.   I. Title.

  LC1768.B8P65 2008
  373.1822—dc22   2007045361

A catalogue record for this book is available from the British Library.

Design by Westchester Book Group.

First edition: June 2008

10 9 8 7 6 5 4 3 2 1

Printed in the United States of America.

For Jon Eben

CONTENTS

ACKNOWLEDGMENTS

I OWE SO MUCH TO THE TREMENDOUS MENTORS I HAVE HAD THROUGHOUT my academic career. This project began as a doctoral dissertation at the University of British Columbia, where I was a graduate student in the interdisciplinary department of Educational Studies. I was fortunate enough to be mentored by two professors who modeled feminist principles not just in their work, but also in their relationships with students. Deirdre Kelly supervised this project in its first incarnation. In knowing Deirdre, I have come to understand the kind of teacher and researcher that I would like to be. I thank her for her superb example, as well as her friendship, sense of fun, and insightful critiques of teen flicks. While at UBC, I was also fortunate enough to work with Dawn Currie in the department of Sociology. It was in Dawn's feminist methodologies class that I wrote my first paper on girls' style in the school, and where I first encountered her unwavering support, intellect, and generosity of spirit. I thank Dawn for the many incredible opportunities she has offered me, as well as her friendship and trust.

I also wish to extend my deepest thanks to Bronwyn Davies, who read several drafts of this project and offered supportive and helpful feedback. Her work on poststructural theory in relation to schooling and gender has influenced me deeply, and I am grateful that she generously agreed to supply the foreword for this book.

Of course, I could not have completed this project without the help of the many caring and smart individuals who offered advice and encouragement. I am very grateful to Jon Eben Field and Hans Skott-Myhre for reading and commenting on drafts of this book. Thanks to Marguerite Pigeon for her artful transcription and uncanny ability to read between the lines. Thanks of Marianna Sirianni for her incredible organization skills and indexing assistance. And thanks to those friends and colleagues with whom I discussed ideas, problems, anxieties, and woes: Amanda Benjamin, Arial Boulet, Cathy Chaput, Nancy Cook, Fanny Dolansky, Rachel Davies, Nicola Doughty, Michael Hall, Sid Katz,

Shannon Moore, Terry Parker, Michelle Pidgeon, Rebecca Raby, Monique Silverman, Kathy Skott-Myhre, Elyssa Katz Tabac, Rachel Toyen, Sarah Twomey, Abby Wener Herlin, Sophie Yendole, and Anne Zavalkoff. Thanks also to my new and supportive colleagues in the department of Child and Youth Studies at Brock University and to the folks at Palgrave, especially Amanda Moon, and Brigitte Shull.

My family has offered me the kind of support that one need never ask for but is always there. Thank you to my parents, Hart and Nancy Pomerantz, for teaching me to enjoy great books, warm bagels, good jokes, and rich conversations about all things big and small. Thanks also to the rest of my family for their humor, friendship, and deeply felt support: Bill Pomerantz, Jennifer Pomerantz, Gail Morgenstern, Erik Goldsilver, Adam Pomerantz, Amanda Pomerantz, Josh Pomerantz, Roger Field, Lorraine Field, Toby Field, Sarah Slipp-Field, and Ellen Field.

I owe a huge shout-out to the many cool and interesting girls I came to know at East Side High. Thank you for your time, patience, stories, and styles. As well, I would like to thank the many teachers who allowed me to sit in on their classes day after day, especially Ms. Mackenzie, who offered me my very first seat in an ESH classroom.

While I owe all of these people my warmest thanks, it is Jon Eben Field who made the writing of this book possible. Not only was he perpetually available to discuss, read, edit, and listen, but he also offered himself up wholly and freely to this process without resentment or complaint. For his generosity of spirit, monster intellect, delicious cooking, excellent taste in music, subtle jokes, sudden interest in style, and unconditional love, I dedicate this book to him.

FOREWORD

THIS BOOK TELLS THE STORY OF A POSTSTRUCTURALIST ETHNOGRAPHY SET in an inner city school in Vancouver. It focuses on girls' dress, analyzing it as "social skin," that is, as integral to the ongoing accomplishment of individual identity, and at the same time, as a system of signs through which those girls are subjected, and with which they must negotiate a recognizable identity for themselves. Ethnographies are, in commonsense terms, stories of "real" places. Understood in such terms, ethnographies entail the researcher gaining the trust of competent informants from within the group being studied. The ethnographer extracts the information that will enable the reader to understand the real world of that group. But in poststructuralist terms, this commonsense realist tale is a misleading one; it produces a fiction and then persuades the reader to believe it; that is, it produces a new authoritative account, which can be made to function as Truth.

A poststructuralist ethnography, in contrast, shows how those fictions, both the researcher's and the participants', are made to work as truths, and in doing so it shows the interplay among discourses, discursive practices and identities, and the shifting relations of power. A poststructuralist ethnography, like this one, takes you into the school, introduces you to the girls in the study, and asks you to put most of your everyday assumptions on hold while you travel with the researcher down the subtle, complex path of coming to see how meanings and identities are negotiated moment by moment, day by day, and how the exchanges among the participants work to make this particular version of human life possible. What is interesting then in this ethnographic study of girls' style is not so much *what* the girls wear—what images they have taken up this year or last year—but how we might understand differently the relations between dress and address, between bodies and the social, and how those relations are played out by girls in the contexts of global capital and of schooling.

This study situates itself in marked contrast to the large number of research studies and media observations in which girls are seen in negative terms, and in which their style of dress is used to demonstrate that lesser status. They are portrayed as ignorant dupes of capitalism, as sexually vulnerable (or sexually available), as mere copycats, and as objects of another's gaze, always positioned in the negative half of a binary in which they are other to the masculine, the rational, the agentic, the innovative.

What this book does so powerfully is to deconstruct the binaries in which girls are always positioned as other to that which is dominant and active, even by feminist researchers. Pomerantz argues that the girls in her study *are* agentic. She shows them taking up an existing sign system, as they must, and making it their own, working with it to accomplish both individual identities and forms of social order. By way of example, she tells a small story against herself. In the early days of her year-long study, she describes herself anxiously trying to figure how to dress appropriately in order to become recognizable as both a researcher and as one who can mix among the girls and be accepted. She was so paralyzed by her anxiety about what to wear that she often missed the bus to school. One of the girls in the study explained to her that it didn't matter what she wore, the thing was how she wore it: "Rock what you're wearing" Gianna told her, with a snap of her fingers, "own it." That vulnerability to the gaze of others and the paralyzing uncertainty that goes with it, and that capacity to live the style as if it grew on your body naturally, are the ambivalent poles of experience that each girl must manage as she finds a way to wear her "social skin."

Everyone, Pomerantz points out, whether male or female, must dress. Every one of us, whether we like it or not, enters into particular forms of recognizability through the clothes we put on. What this book does so well is to show the necessary doubleness of these acts of recognition. One becomes recognizable through what Butler calls "circuits of recognition." Each one of us must become recognizable within those circuits, or be cast out. In order to be addressed at all, acts of dressing must first locate the subject within the already existing sign system, as a recognizable actor:

> We may think that to be addressed one must first be recognized, but [I suggest] the address constitutes a being within the possible circuit of recognition and, accordingly, outside of it, in abjection. . . . One "exists" not only by virtue of being recognized, but, in a prior sense, by being *recognizable.* The terms that facilitate recognition are themselves conventional, the effects and instruments of a social ritual that decide, often through exclusion and violence, the linguistic conditions of survivable subjects. (Butler 1997a, 5)

In this sense, styles of dress are necessary for the accomplishment of social status, of recognizability, of a position in the world that saves one from being cast out. Dress is a system of signs, a form of language, that we both depend on, and that enables us to become autonomous subjects, however illusory that autonomy may be. The prior existence of that language, as Butler points out, is a kind of insult. We are vulnerable to its power both to shape us and potentially to harm us. She asks in the context of her study of violent speech acts:

> Could language injure us if we were not, in some sense, linguistic beings, beings who require language in order to be? Is our vulnerability to language a consequence of our being constituted within its terms? If we are formed in language, then that formative power precedes and conditions any decision we might make about it, insulting us from the start, as it were, by its prior power. (Butler 1997a, 1–2)

Pomerantz carefully sidesteps the argument that might follow from this construction of dress as a sign system with "formative power [that] precedes and conditions any decision we might make" by following carefully the poststructuralist argument about discourse:

> Discourse is not merely spoken words, but a notion of signification which concerns not merely how it is that certain signifiers come to mean what they mean, but how certain discursive forms articulate objects and subjects in their intelligibility. In this sense "discourse" is not used in the ordinary sense. . . . Discourse does not merely represent or report on pregiven practices and relations, but it enters into their articulation and is, in that sense, productive. (Butler 1995, 138)

In the context of this study, dress, as a mode of signification, articulates subjects; it forms both intelligibility and individual subjects as intelligible. Through dress, through "making dress rock" by "owning" it, subjects form their own location on the grid of intelligibility. They grow their social skin. They take up a viable life through becoming recognizable. They become *this* and not *that*. They are located *here* and not *there*. They perform things with the sign system of dress—not only particular identities, but the social order inside of which those identities can be made to make sense. The focus here is not on what we wear, as Gianna points out, but what is accomplished in that wearing. It is the performative power that is of interest:

> We do things with language, produce effects with language, and we do things to language, but language is also the thing that we do. Language

is a name for our doing: both "what" we do (the name for the action we characteristically perform) and that which we effect, *the act and its consequences*. (Butler 1997a, 8, emphasis added)

What this theorizing does is to unhook girls from the more ordinary sense of performance (a superficial play at being someone or something else) and makes sense of their style as performative: They accomplish themselves as recognizable social beings with social presence within a recognizable social order. The power of the market acts on them, it provides them with the tools they need to become recognizable, but through their own actions that power becomes their own as they take it up through their own practices: "Power acts on the subject in at least two ways: first, as what makes the subject possible, the condition of its possibility and its formative occasion, and second, as what is taken up and reiterated in the subject's 'own' acting" (Butler 1997b, 14).

The meaning of a particular style is not in the object/body/dress/ social skin, but continually negotiated within complex social relations. In the daily lives of the girls the readings they produce of each other are always potentially slippery. A girl who tries out a popular look and doesn't manage to own it, or is not seen to manage to own it, might be seen as trying too hard, and as such, worthy only of contempt. The loud popular girls might comment of a girl walking by saying, "Oh my god! That girl's trying to look like Avril Lavigne!" Or just, "What are they thinking!" A girl who develops her own very particular look unlike anyone else might impress others with her capacity to maintain an "alternative" style, and at the same time find very few willing to befriend her. Girls who adopted a preppy style might be viewed by others as "skanky." In Gianna's words again "There's a very close line between looking good and looking like a ho." In other words, the work the signs might accomplish both depends on what went before, that is, what meaning making went before, and also on how it is made to live in any one present moment and context and set of relations on the body of a girl—not just by the way she wears it, but by how she is recognized by others as wearing it. It is a dynamic interplay of power among girls, between girls and the global marketplace, and between the present and the past.

Pomerantz strikes a fine balance throughout this book between making visible the considerable impact of external forces on the girls in her study, and arguing the poststructuralist position that the girls have agency. This is not the agency of the sovereign subject, but the agency of the reflexive, ambivalent subject, the one who sees the forces at work on her, and who works with and against those forces to make not only a place for herself in a pre-existing discursive universe, but in the act of making that discursive system her own, using the system to act back on itself. These girls are branded by the market. As one girl says, the images and the desire for those images are "imprinted on your brain . . . like a tattoo that you can't get off." And yet this flow of power is not all one way, the market must listen to the girls as they change their minds about what they will wear and about what particular clothes and brands will come to signify. The market can work hard to insert particular meanings and desires, but it is within the girls' power to take them up, resist them and sometimes abandon them.

This is an engaging book. It makes an important contribution to the growing body of poststructural ethnographies, and it makes a significant contribution to thinking about girls' identities within social relations, discursive networks and the symbolic economy of dress.

Bronwyn Davies
Narrative Discourse and Pedagogy
University of Western Sydney
2007

REFERENCES

Butler, Judith. 1995. For a careful reading. In *Feminist contentions: A philosophical exchange,* ed. S. Benhabib, J. Butler, D. Cornell and N. Fraser. New York: Routledge.

————. 1997a. *Excitable speech: A politics of the performative.* New York: Routledge.

————. 1997b. *The psychic life of power: Theories in subjection.* Stanford, CA: Stanford University Press.

CHAPTER 1

BEYOND THE "SLUT" LOOK: FROM INCITEMENT TO IDENTITY

Yeah, the low jeans are very cool. And you gotta show your thong.
Ratch, "punk" style

I WAS IN A DOWNTOWN COFFEE SHOP THE FIRST TIME I RECOGNIZED IT AS a style. I had just settled into my seat when I looked up to see the "ass crack" of the teenage girl sitting in front of me. Her jeans were so low on her hips that they simply could not contain her backside. I sat for a moment wondering if I should tell her. Would she want to know? After debating the issue in my head, I walked over to where she was sitting and politely interrupted the conversation she was having with a friend. "Yes," she murmured, looking up at me. Suddenly realizing the bizarre nature of the news I was about to deliver, I leaned in closer so as not to embarrass her. "Um, I was just sitting behind you, over there, and I noticed that, well, I could see your, um, I could see your behind. Like, *quite* a lot of it. I just thought you might want to know." I figured that she would say a quick "thanks" and then head to the bathroom to pull up her pants. But she did not move. Instead, she stared awkwardly at her friend. "Okay, sorry to bother you," I said quietly, suddenly realizing that the message I had been at pains to deliver was unappreciated. I started to back away. "I-I just thought you'd want to know. I mean *I* would want to know, so I thought you would too." I went back to my table, somewhat confused. When I felt compelled to look again, I saw the girl, who might have been sixteen or seventeen, reach around and touch the back of her jeans. Was she checking to see if I was right, that her jeans were indeed embarrassingly revealing? She gave them a quick tug, bringing the waistband up about a quarter of an inch. But the rest, I suddenly realized, was meant to show. Turning to look at me, both she and her friend giggled. I was just an adult who could not possibly understand.

After that, I saw the style everywhere. In truth, it was impossible to miss. Anyone who rode the bus or spent time in malls or knew any young people at all would have seen it. Waistbands had descended. Shirt lengths had risen. Necklines had plummeted. Bra straps had emerged. Breasts had become semi-exposed. Jeans had become "painted-on" tight and low on the hips. Material had become clingy. Body parts had become uncovered. Underwear had become visible. To those who were not sporting this style, it seemed strange and audacious and perplexing. I could hear people, mostly adults, speaking about it as teenage girls passed by. I watched them follow girls with their eyes, make a face, and turn to their friends and say, "Did you just see that?!" Most often, I heard people use the word *disgusting* to describe the look. "That's *disgusting*!" they would say. Other times, I would hear the word *slutty,* or its new synonym, *skanky,* used to describe what they just saw. "That's *so* slutty looking!" or "*What* a skank!" These were the things that people who were not wearing such items of apparel said. But for those who were, there seemed to be no trace of awkwardness, no hint of concern. It was just business as usual. A trend. A look. A style.

While I was indeed shocked the first time I noticed the style, this book is not a condemnation of how girls are dressing "these days." Such judgments are plentiful enough in the panic-driven media and everyday talk of adults who claim to know what is best for girls. Instead, this book takes a different approach by pushing beyond those judgments in order to understand girls' style as a "*deep* surface" (Warwick and Cavallaro 1998, xxiii). While style sits on the surface of girls' bodies, it is anything but superficial. As Elizabeth Wilson (1985) humorously notes: "For everyone, clothing is compulsory." For girls, this statement is especially true. Not only is clothing compulsory for girls, but its careful deliberation is all but mandatory in the school, where "how you look" is often equated with "who you are."

This book seeks to attend to the complexity of girls' style by viewing it as a powerful form of identity construction and negotiation. How do girls use style to engage with intersecting identity categories, such as gender, race, ethnicity, class, and sexuality? How do girls use style to insert themselves into a social world—as individuals and as members of groups? How do girls use style to signal belonging, friendship, politics, resistance, ambivalence, anger, lifestyle affiliations, individuality, image, and personal taste?

I offer no pronouncements on whether or not girls "should" dress the way they do or whether or not particular styles are "better" than others. Instead, I hope this book takes readers deeper into girls' cultural practices by exploring the contextual significance of girls' style as a mode of self-expression, identification, and agency. Given this framing, I view girls' style as more than "just fashion." Style is a major part of how girls understand themselves and each other, offering them something to cultivate as their very own—whether they do so by making their clothes, in the spirit of DIY (do-it-yourself), or by buying their clothes at the mall, in order to keep up with the trends. This book is thus grounded in the simple assertion that girls' style is important not just to girls themselves, but also to anyone who might take an interest in girls and girlhood; not as a "problem" in need of solving (the *message du jour* in the academic and popular press), but as a complex and fascinating social category deserving of attention.

Girls' style is made up of an unlimited array of possibilities, encompassing anything that embellishes, covers, or decorates girls' bodies. Articles of clothing, such as jeans, hoodies, tank tops, T-shirts, dresses, skirts, undergarments, socks, and shoes are a part of girls' style. Accessories, such as jewelry, belts, hats, scarves, backpacks, purses, buttons, broaches, and bags are a part of girls' style. Bodily adornments, such as hair color, nail polish, makeup, piercings, tattoos, and drawings etched on the flesh are a part of girls' style. Accoutrement, such as iPods, headphones, cell phones, discmen, cigarettes, and even books and CDs that are carried around in order to be seen are a part of girls' style. Beyond these individual items, style is the way in which all of these elements interact with one another and come together to create a girl's overall look. A girl's look is the effect she achieves through her style—be it punk, goth, skater, hip hop, b-girl, gang, rave, preppy, sporty, club, skid, metal, grunge, straight-edge, hippy, hipster, retro, random, dressy, alternative, mainstream—or expressions she uses to describe her style, such as "pretty normal," "just regular," "totally average," or "no real style at all."

But this broad definition is only the tip of the iceberg. Girls' style has always been a touchstone for social sensibilities in North American society, acting as a litmus test for shifting cultural values and norms. Girls' style has pushed the boundaries of "normal" and "abnormal," "respectable" and "risqué." Through style, we can trace women's liberation and attempts to resist and expand gender norms, standards of feminine beauty, and the carefully patrolled borders of the heterosexual body. But we can also trace moral outrage, bylaws, and dress codes designed to keep girls and women in line. As Kelly Schrum (2004) notes in her historical analysis of girls' culture from the 1920s to the 1940s, style can act as a physical manifestation of dominant ideologies, used to police "acceptable" expressions of femininity, sexuality, autonomy, and pleasure. Style is thus a form of disciplinary power that keeps girls' bodies under wraps and under control—a tool of surveillance that enables others to monitor what girls are up to, thereby compelling girls to monitor themselves.

But just as girls have been policed through style, so too have they used it to move girlhood into uncharted territory. In the 1930s, for example, girls started wearing pants to school for the first time, inciting debate over appropriate female attire. Schrum notes that high school boys expressed uneasiness "with the shifting boundaries of gendered clothing, perceiving a threat to their emerging masculinity by the erosion of fashion categories that designated pants as 'male'" (55). In the 1930s, short shorts also appeared on the scene, creating panic in New York City that led to an ordinance banning this "offensive" item of apparel. But many girls ignored these laws, and "highly publicized efforts to control the clothing of young women in public places proved futile" (57). By the 1950s, girls' dress had driven a wedge between generations, solidifying "teenagehood" as a separate life stage. In the 1960s, girls wore sexually liberated, countercultural styles, pushing that wedge to its limits by representing resistance to authority and the dominant symbols of middle-class life. In the 1970s, girls dabbled in androgyny, as well as the aggressive styles of punk rock and new wave. Girls' style also became branded with a naughty yet innocent sexuality. As Brooke Shields cooed, "You know what comes between me and my Calvins? Nothing." In the 1980s, the new medium of music videos catapulted girls' style into the flashy world of pop star chic, where MTV helped to popularize "raunchy" Madonna-inspired clothing, including underwear as outerwear, religious icons as ironic accessories, lacey bustiers, and tummy-revealing "pop tops."

And now, at the dawn of a new century, girls' style continues to test the limits of acceptable bodily display and sexual propriety. In the 1990s and early 2000s, manufacturers, marketers, and retailers began promoting a style based on the massive popularity of pop star Britney Spears. In the "Britney" look, girls sometimes revealed the V-shape of their groins, their "ass cracks"—also known as "plumber's butt"—and, in extreme cases, the top part of a well-waxed or closely shaved pubic area. Navels—usually adorned with piercings—were observable in midriff-revealing shirts. And girls' breasts were often only half concealed in tank tops and T-shirts made of thin, clingy material with long spaghetti straps that emphasized a girl's cleavage. Though shockingly sexual and down-right scanty, such items of apparel flew off the shelves in malls across North America. But the style was not at all popular with many adults, who saw nothing but trouble in the "slut" look, as it also came to be known, for its perceived proximity to what many thought of as "hooker" wear.

The amount of print space and airtime devoted to the "slut" look was staggering. Many talk shows, newspapers, magazines, and radio pro-grams dedicated themselves to "covering" the issue, calling in "experts" to field questions from concerned parents. "If my daughter dresses like a 'slut', does that mean she is having sex?" "Should I let my daughter out of the house looking like that?" "Is my daughter going to be sexu-ally assaulted if she wears low-rise jeans?" The issue of girls' style had become an incitement to discourse, though certainly not for the first (or last) time. As Michel Foucault (1978, 8) explains, the incitement to discourse over sex from the eighteenth century onward was produced by "a determination on the part of the agencies of power to hear it spo-ken about, and to cause it to speak through explicit articulation and endlessly accumulated detail." In the case of girls' style, a similar deter-mination existed on the part of the media and the general public to sustain negative attention over what girls were wearing "these days." More than just fashion news, the "slut" look occasionally made it to the front page, announced in towering headlines and colorful images that would have struck fear in any parent's heart. The articles usually came in three varieties: concern for girls' well-being, allegations that girls were "asking for it," and accusations that girls were engaging in im-moral behavior.

In this book, I offer another perspective on girls' style, one that is not commonly expressed in the media or by the general public. I wanted to link girls' style to a specific context, namely the school. Style, like any cultural practice, only makes sense when viewed in rela-tion to the social codes and locales that grant it meaning. But reporters rarely have the print space or time to devote to understanding those contexts. Instead, stories must be pulled together in a matter of days or hours, without the luxury of attending high school for a school year in order to make connections between cultural texts and social relations. There is no time to observe girls in classes, to conduct in-depth inter-views, or to try to understand the overt and covert ways that girls might use style to negotiate their fluid school identities. But as a femi-nist ethnographer, I had the time to explore these issues—issues that I felt had wider ramifications for how our society looks at, treats, and val-ues girls. I wanted to focus on style in relation to girls' lives, particularly how girls used style to make themselves visible within their school's social world. I wanted to learn how girls saw themselves and others as a result of style, and how style related to intersecting identity categories, such as gender, race, ethnicity, class, and sexuality. In short, I was interested in the story that did not appear in the press or surface during the many animated conversations I overheard between adults as teenage girls passed by.

There are no fixed portraits of girls in the stories that follow. I have not determined once and for all what styles "mean" and why girls wear particular items of apparel. Such single certainties are not only impossible, but also undesirable. Instead, I have tried to offer a view of girlhood that I felt was missing not only in the press, but also in the public imaginary—a view that I hope adds a layer of complexity to the customary narratives surrounding girls' cultural practices, as well as girlhood itself. This book thus seeks to move beyond the "slut" look by focusing on style as girls used it to craft images, identifications, and affiliations in the school. But in order to get to those stories, other stories must first unfold. And so it is here, in the midst of a "hot topic," that my exploration into girls' identities begins.

BRITNEY SPEARS, SCHOOLGIRLS, AND SINGLE CERTAINTIES

I began clipping articles on girls' style from newspapers and magazines in 1999. Not coincidentally, it was also the year that Britney Spears hit number one on the Billboard chart with her debut album, *Baby One More Time* (Jive Records). I clearly remember the first time I saw the video for the title song. My roommate called out to me from the living room. "Shauna, you have to see this!" I came running to the television only to discover that I had been called in to watch a pretty teenager singing a fairly average pop tune. But something about the video was unique. It combined the sweet innocence of schoolgirl naivety with the dirty sexuality of a striptease. While such imagery was rampant in the videos of male rock bands, it was uncommon—even unheard of—for a seventeen-year-old (who claimed to be a virgin) to use such a routine en route to megastardom.

The controversial video featured Britney in a schoolgirl uniform, but with a unique twist. At her own suggestion (Daly 1999), she knotted the white blouse she was wearing just under her breasts to reveal her midriff. She then proceeded to dance down the hallway into infamy.[1] Though Britney was certainly not the first pop star to expose her midriff (she owes much to Madonna and the Spice Girls for this trademark, not to mention girl groups of the 1960s, 1970s, and 1980s), her name became associated with a new kind of style that had recently invaded malls and classrooms across North America. As one reporter wrote:

> Blame it on Britney. As teens across the country negotiate the sexually charged territory of high school, more and more of them are doing so in cleavage-baring crop tops and ultra-low-rise pants that show off their pelvic bones. Clothes for girls have never been so skimpy. (Fulsang 2002, L1)

The look in question included low-rise jeans, midriff-revealing shirts that showcased pierced navels, visible bra straps, tight tank tops, micro miniskirts, underwear that "peaked" above the waistband (particularly the "whale tale" of thongs and g-strings), and other items of apparel that either exposed a girl's body or made it highly visible through an extremely tight fit. Such clothing caused the *Washington Post* to bemoan the "Whore Wars" (Stepp 2002), the *New York Times* to denounce "the teenage strumpet look" (Trebay 2003), and the word *slut* to regularly appear alongside images of teenage girls in newspapers and magazine articles. *MacLean's* (George 2007) magazine asked, "Why are we dressing our daughters like this?" and *Newsweek* (Deveny et al. 2007) wondered about the "harmful" influence that Britney's style has had on girls in the cover story, "Girls Gone Bad?" Columnists, too, eager to embrace this "hot" topic, relied on the new urban slang for eye-catching headlines: "Prostitots," "Hoochie Mamas," "Kinderwhores," "Hos," "Lolitas," "Teleslutties."

While print articles condemned the "slut" look, editors were only too happy to select images of girls clad in "offensive" outfits to accompany the text. Girls were routinely posed in suggestive clothes that made them look wanton and immoral. One such photograph featured a full-page color image of a very skinny teenage girl outfitted in red fishnet stockings, cut-off jean shorts, a purple bikini top with silver glitter, and a large heart-shaped tattoo between her breasts. Her face was hidden by a mass of hair that made her look wild in her wayward attempt to get out of the house in the outfit no matter what the cost to her relationship with her family or her reputation. The accompanying headline read, "Mom, I'm ready for school!" (Fulsang 2002, L1). Articles and images such as these nailed the "problem" down as one of indecency, risk, and morality. Parents felt they knew "best," while teenage girls were cast as the unconscious dupes who did not understand the consequences of their actions. As Foucault (1978, 20) notes in regard to the rise in talk surrounding sex, the issue was thus "taken charge of, tracked down as it were, by a discourse that aimed to allow it no obscurity, no respite."

By the early 2000s, the Britney look was everywhere, setting off the latest wave of moral panic to surround girls' style and, thus, girls' behavior.[2]

While the panic rippled across various locales, it was inside the school where the issue reached its boiling point. When girls began arriving to class in low-rise jeans and midriff-revealing tops, teachers and school administrators were caught off guard. How would students learn in such a climate? How would boys and teachers be able to concentrate?[3] How would girls be able to sit down without revealing their "ass cracks" to those behind them? The "slut" look quite literally threw the public school system for a loop, causing debates over the purpose of the school. Was the school for strictly pedagogical purposes, in other words, teaching and learning? Or was it a social arena for the experimentation of teenage sexuality? The answer revealed (and not for the first time) a shocking double standard between the treatment of boys and girls. While the display and enactment of male sexuality has never been questioned as inappropriate in the school, the same consideration was not extended to girls. This double standard exemplifies what Michelle Fine (1988) calls the "missing discourse of desire." While boys' "raging" hormones are seen to be "unavoidable," the discourse surrounding girls' sexuality is virtually nonexistent. Instead, girls are cast in the role of gatekeepers of adolescent male sexuality, charged with the task of curtailing the "rampant" and "uncontrollable" urges of boys in order to maintain order and morality in the school (Fine 1988; Kenway and Willis 1998; Pomerantz 2007; Tolman and Higgins 1994). While "boys will be boys," girls must control themselves, as well as the young men around them.

In order to deal with the "situation," calls for a prohibition on girls' bodies were fast and furious, as administrators scrambled to update "inadequate" dress codes for the new millennium. In order to prevent girls from attending class in outfits that would certainly distract boys from their work and teachers from their teaching, principals feverishly rewrote school dress codes, many of which had not been altered for decades.[4] Midriffs, cleavage, skirts above a certain length, low-rise jeans, visible bra straps and underwear, and even tank tops with spaghetti straps were banned from many North American schools. Principals kept long T-shirts and baggy sweaters in their offices for quick cover-ups, sometimes suspending girls who came to school wearing anything "inappropriate" (Pomerantz 2007). Under the guise of social order, a twenty-first-century dress code emerged that continued to place the burden of self-control, public decency, and sexual morality in the school on girls' shoulders.

Aside from placing girls' bodies under surveillance and the ridiculous assumption that covering them up was tantamount to solving the problem of sexual harassment or curbing female sexual desire in the school, other problems existed in the moral panic surrounding girls' style. Only one version of female youth was represented in accounts offered by the press, creating a standardized form of girlhood that came in one shape, size, and outfit. In his study of written fashion, Roland Barthes (1983) demonstrates the power of carefully organized and constantly repeated images. Through such repetition, a "single certainty" is fostered that separates meaning from its social, cultural, and historical antecedents (13). Images are isolated from those who created them and the cultural context in which they were produced, causing them to take on a life of their own that seems inevitable—an always-already stance, posture, and style that ceases to be out of place given its constant repetition. Dorothy Smith (1988, 37) observes a similar inevitability in the repetition of images in advertisements aimed at female consumers, where femininity is seen as a "determinate and unitary phenomenon," rather than as a random collection of events made to look whole. As Dick Hebdige (1979, 85) observes in relation to youth, through such images, girls are "made to cohere in contradiction," where one girl in low-rise jeans and a visible thong stands for *all* girls, no matter how many variations, adaptations, and deviations there might be.

The "single certainty" in the press offered the public a very literal reading of girls' style: This is what a slut looks like. But as Tony Jefferson (1975, 55) explains, cultural objects can have no singular meaning. "They 'mean' only because they have already been arranged, according to social use, into cultural codes of meaning, which assign meaning to them." Similarly, in her study of teen magazines, Dawn Currie (1999) warns against analyzing cultural objects as stable facts, an assumption that makes it seem possible to read the "truth" about girls off their bodies. I encountered such decontextualized judgments numerous times when people would offer me their "readings" of girls' style based on casual observations and quick glances. "Girls must have low self-esteem to dress like that," many said, or "Girls only dress like that to attract boys." But Currie challenges us to transcend the idea that cultural objects offer self-contained units of "truth." The question, then, is not, what do these practices mean? But why do these practices have meaning for girls in a specific place, during a specific time in history?

SYMBOLIC ECONOMIES OF STYLE

Long after I completed the research for this book, newspaper articles, magazine stories, and columns were still being written about girls and their "slutty" styles. As I clipped them out for insertion into an already overflowing file folder, I thought of the variety of girls I had come to know during my year at East Side High (ESH), the urban and multicultural high school in Vancouver's east side where I conducted the research for this book. During the 2002–2003 school year, it was widely ac-knowledged that there were two particular "uniforms" for girls—what girls often referred to as the Britney and JLo looks. The Britney look was synonymous with the "slut" look discussed above, but at ESH, it took on more complicated significations than the press or the public might have imagined. In one English class, I was struck by a group of thin, white girls in the back corner who sported the Britney look. At first, they all seemed to be dressed the same: tight, low-rise jeans, sneakers, midriff-revealing T-shirts or tank tops, square silvery earrings, glossed lips, and long, blonde hair. But over the course of the year, I came to see the subtle markers of distinction that existed among them, such as who could afford to buy "labels"; who had the freedom at home to dress any way they wanted; who was considered to be "tom-boyish"; and who was con-sidered to be "too" sexy.

There was also the JLo look, so-called for its Jennifer Lopez influ-ence, worn by many Asian, Italian, Hispanic, and, to a lesser extent, white girls. A matching tracksuit made of velour defined the style. The "suits," as they were called, had low-rise bottoms that were considered sexy because of the way they "hugged" a girl's figure. But this uniform was also not homogeneous. Some girls wore name brand suits from *Aritzia* (approximately $150 for the matching set), a "high end" casual store in malls that caters to teenage girls, while others wore "no name," knock-off versions, reflecting a girl's class location. In short, not all uni-forms in the school were created equally and each seemingly categorical style was open to countless significations and interpretations based on who was doing the looking and who was being looked at.

Aside from these two acknowledged uniforms, girls at ESH also de-scribed their styles as comfortable, sporty, goth, punk, alternative, dressy, classy, preppy, regular, casual, weird, skater, random, hip hop, and vari-ous combinations and hybridizations of each. Even within these already fluid categories, distinctions across racial and ethnic groups, social classes, and lifestyle affiliations overtly and covertly marked girls. A pair of regu-larly worn striped socks brought Abby fame within the school. Admit-tedly unpopular, her trademark socks put her on the map. As she put it, people would say, " 'Uh, who's Abby again?' 'Um, yeah, the girl with the striped socks.' " More to the extreme, Ratch, a punk, cultivated a look that separated her from almost everyone at ESH. One day she walked into English class wearing an orange crinoline tutu that she had picked up at Value Village, a mecca for secondhand shoppers. The rest of the outfit was characteristic for Ratch: ripped fishnets, black converse sneakers covered in cryptic writing, and a Sex Pistols T-shirt. The tutu, however, set her apart from her usual look, which already set her apart in the school. And there was Xiu, a self-described "quiet Chinese girl." Occasionally working to neutralize her studious image, she would wear black in order to give her style "a little kick," a distinction that would only have been recognizable among her group of friends.

In her study of identity among Mexican American and white high school girls, Julie Bettie (2003, 61–62) describes a "symbolic economy of style" in the school:

> A whole array of gender-specific commodities were used as markers of distinction among different groups of girls, who performed race-class-specific versions of femininity. Hairstyles, clothes, shoes, and the colors of lipstick, lip liner, and nail polish, in particular, were key markers in the symbolic economy that were employed to express group membership as the body became a resource and a site on which difference was inscribed.

The press and the public rarely acknowledge that styles correlate to a symbolic economy that shifts from school to school, or that styles carry markers of distinction among racial and ethnic groups, or that styles—particularly the ability to afford labels—carry distinctions between middle-class, working-class, and working-poor girls.[5] These contexts are often ignored in the snap judgments made about girls, even though they are the very "stuff" from which girls make "forced choices" (Davies 1990, 46) about how they want to look in the school.[6]

This book aims to situate girls' style within the context of ESH's social world. It also aims to reconceptualize style as a significant form of identity construction and negotiation within the school, where it becomes a "particularly powerful social marker" (Eckert 1989, 62). As Catharine Driscoll (2002, 245) suggests, while style is "subject to regulation (school roles and public laws, for example) and limitation (available finances, for example) . . . it is also always an articulation of girls' cultural identities." Such articulations get very little airplay in the panic-driven media or among concerned parents, teachers, and school administrators, who see "sexy" clothing as an indicator of danger and disruption. But beyond the "slut" look and its surface level readings, there lies a complex social world in which girls negotiate their identities using, among other things, style.

THE DOUBLENESS OF IDENTITY

I have chosen to use the word *identity* because, as Hall (1996) suggests, it is still a meaningful and necessary concept. But Hall also reminds us that if identity is to be considered at all, it should be considered under *erasure,*[7] hovering, as it does, "between reversal and emergence; an idea that cannot be thought in the old way, but without which certain key questions cannot be thought at all" (2). Placing "identity" under *erasure* is an acknowledgment that the word is problematic, loaded with thorny meanings that are no longer relevant. But it also points to the fact that the word is not yet obsolete, offering a valuable way to theorize the relationship between self, other, and society. Interestingly, no matter how complicated the concept of identity has become, it still appears regularly in academic work, particularly in ethnographic studies done in high schools (see Bettie 2003; Gonick 2003; Eckert 1989; Yon 2000) and anthologies dedicated to the experiences and cultural practices of girls (see Bettis and Adams 2005; Harris 2004a; Mazzarella and Pecora 1999). This continued usage, particularly in research on young people, suggests that identity is still a powerful and desirable concept with which to think through the making of the self, however messy and incongruous that making may be.

In placing identity under erasure, I am indicating that it is, to use Gayatri Spivak's (1974) word, *inaccurate.* This inaccuracy stems from its humanist roots as a stable category that defines an individual's essence or core being. Through this lens, identity is the "Self" with a capital "S," meaning a "conscious, stable, unified, rational, coherent, knowing, autonomous, and ahistoric individual" (St. Pierre 2000, 500). Identity thus becomes our most enduring feature, stemming from an innate sense of being that makes us who we are. This core promises what Jacques Derrida (1978) calls full presence, or an indissolubility that makes us complete—an unwavering, unchanging origin that precedes language and society. Through this lens, identity categories are also stable and fixed. Race, ethnicity, gender, class, and sexuality are viewed as inborn characteristics that fix the Self as, say, female, Chinese, working-class, heterosexual, and first generation Canadian. Such categories are not seen to be fluid or negotiable, but mark the Self with indelible qualities that endure over time and space, fixing identity in one coherent and knowable package.

But this humanist view does not take into account all the ways in which we are acted upon by external forces and, instead, presumes that identity is a solitary and internal event that escapes the weight of history and the ubiquity of power relations in society. Hall's (1996) belief that identity is still a useful concept, then, comes from its radical resignification within poststructuralism, where the subject is far from complete and stable. Instead, the self is always partial and unfinished, contingently forming and reforming in relation to others, social structures, and our own multiple and contradictory subject positionings. This shift, as

Davies (1997, 274) explains, "entails a move from the self as a noun (and thus stable and relatively fixed) to the self as a verb, always in process, taking its shape in and through the discursive possibilities through which selves are made." The poststructural subject is thus a "linguistic category, a place-holder, a structure in formation" (Butler 1997, 10), rather than a stable entity understood through internal "mastery and control" (Smith 1987, 66).

The instability of the subject occurs through its constitution in discourse, or the way in which language, meaning, and thought are organized by institutions and sociocultural forces that have the power to delimit "right" and "wrong," "normal" and "abnormal," "beauty" and "abjection." Such forces include education, government, medicine, religion, law, family, media, psychology/therapy, academic and professional research, and all the ways in which the rules and norms of these solidifying institutions become disseminated, digested, and internalized as "truth." Through the regulatory and disciplinary power of these forces, particular ways of knowing and being are venerated and condemned. Certain modes of girlhood, for example, are held up as "good" and "natural" while others are classified as "bad" and "deviant." Discourse has the authority to solidify socially constructed categories, granting them authorizing power that forms commonsense ways of thinking about others, the social world, and ourselves. Our identities are thus shaped by discourse, as it defines both self and society for us.

But while the poststructural subject is constituted through discourse, it is not wholly determined by it. Instead, as Hall (1996, 5–6) suggests, identity is the

> meeting point, the point of *suture*, between on the one hand the discourses and practices which attempt to "interpellate," speak to us or hail us into place as social subjects of particular discourses, and on the other hand, the processes which produce subjectivities, which construct us as subjects which can be "spoken."

While discourse speaks us into subjecthood—naming and classifying us as particular kinds of people who lead particular kinds of (sanctioned, disparaged, privileged, oppressed) lives—it simultaneously offers us attachment to the social world through the subject positions or social roles that we occupy as a result of our discursive constitution. These attachments grant us stability and a sense of coherence that feels "real" and permanent, but are actually a contextual and temporal sense of self that shifts over time and space (Hall 1990; Yon 2000). As Hall (1996, 6) explains, the result of "a successful articulation or 'chaining' of the subject into the flow of discourse" produces not just interpellation, where the subject becomes that which she is named, but also a subject who is invested in the subject positions to which she has been sutured. This investment is marked by a "passion for identity" (Yon 2000, 1), which causes us to protect, negotiate, reconstruct, challenge, and resist the ways in which we have been positioned—to "speak" back to how we have been "spoken" (Davies 1997).

Being able to "speak" as a subject does not, however, enable us to escape our discursive positioning, nor does it mean we precede discourse as a "doer behind the deed" (Butler 1993). Instead, the suturing of discourse and subjectivity produces a fluid and multiple subject who is both enabled and constrained by discourse at the same time. In calling attention to this fluidity, Judith Butler (1990, 1993) refers to identity as performative. Though the word conjures images of a rational, choosing Self, Butler is quick to point out that performativity is not "theatrical," where individuals "perform" themselves on the public stage of society (see Goffman 1959). Performativity does not denote a "choosing" individual who freely moves in and out of identity categories, like a character in a play who changes costumes in order to signify different modes of being. Rather, identity is performative in that discourse positions us to play particular social roles that are historically, socially, and culturally bound. These roles are constraining in their limitations and their apparent fixity. But they also open up the possibility of reiteration, or a refusal to repeat the behaviors that give such social roles stability and coherence. As a result, the subject becomes "the permanent possibility of a certain resignifying process" (Butler 1992, 13). It is through this resignifying process that subjectivities are produced, nurtured, and creatively used to offset the limiting features of discursive constitution.

Viewed through this lens, identity is fluid, open, and incomplete, produced within the specificity of time and space, discourse and subjectivity, self and society. We co- and reconstruct our identities through identificatory processes that occur in relation to, and not independently of, the social world. It is this understanding of identity that Hall (1996) suggests is worth holding onto, making the word relevant in its radically resignified form. Using the word *identity* thus enables me to discuss both the fixity and the fluidity of girls' identity talk, capturing the doubleness of discourse and subjectivity. While identity is a contingent process, girls also explained themselves in stable and knowable ways that cannot be discounted (Bettie 2003).

On the one hand, girls spoke of themselves as fluid in that they articulated the possibilities that an understanding of discursive positioning could offer. They felt it was possible to negotiate how others saw them in relation to their gendered, raced, and classed subject positions. They articulated their identities as contextual, bound by a specific time and place in history. Girls understood that the subject positions to which they were tethered in the school could and did shift as they migrated to different locales and participated in different social worlds. Jamie, a girl who saw herself in particular ways inside of the school, eloquently described this contingency.

> At school you put on a certain hat. You *are* a certain way. You dress a certain way. You're spoken to a certain way. You speak to other people a certain way, because that's the way you are at *school*. At home, you're very different.

For Jamie, school was the site of a particular performance of identity, one based on her discursive positioning as a white, working-class, heterosexual, fifteen-year-old girl in the tenth grade who embodied an "alternative" ethos and style. These subject positions granted her subjecthood within ESH's social world. But she also recognized that this identity would shift based on how she was positioned within other contexts, in this case, her home.

But on the other hand, girls also felt their identities as fixed in the school. Because identity categories hold specific penalties and rewards, oppressions and privileges, they caused girls to feel stereotyped and defeated in ways that did not feel fluid in the least. Instead, girls spoke of feeling "trapped" by their gender, race, ethnicity, class, and sexuality, enabling some girls to subjugate others, erecting hierarchies and solidifying social boundaries that caused girls pain and anguish. As Bettie (2003, 53) found, while identity categories are theorized as social constructions, they still create "a temporal fixity, bound by the context of history and culture" that makes them feel stable. Theorizing race as a social construction used to justify white supremacy rather than any actual correlation between physical attributes and personality does not negate the racism that girls endured as a result of their racial and ethnic positionings within the school (Bettie 2003; Yon 2000). Theorizing gender as a set of constantly repeated acts within a regulative mold does not negate sexism or the pressure that girls felt to perform girlhood in heteronormative ways (Butler 1990, 1993). And theorizing class as a set of embodied dispositions based on one's access to cultural and economic capital does not negate the pain girls expressed for not being able to afford the kind of clothing that would have made them popular within the school's symbolic economy of style (Bourdieu 1977, 1984, 1998).

Given these considerations, I use the word *identity* in order to explore all of its connotations. While I theorize identity in its shifting and performative sense, I also wish to represent both the fluid *and* the fixed flavor of girls' identity talk. To that end, I use the expression "school identities" in order to attend to this doubleness. Girls saw their identities in the school as fixed in that they were positioned through the constraints of discourse that felt beyond their control; but girls also felt their identities in the school as fluid in that they understood that they could—in recognizing these positionings—negotiate how they were seen by others. This doubleness was unmistakable throughout my ethnographic explorations of girls' experiences in the school, where girls were compelled to perform girlhood in certain ways *and* felt liberated by the possibilities that the socially constructed category "girl" opened up. In order to further contextualize this doubleness, I next explore the notion of subjectivity, particularly as it is embodied through style as "social skin."

STYLE AS SOCIAL SKIN

Chris Weedon (1987, 32) defines subjectivity as "the conscious and unconscious thoughts and emotions of the individual, her sense of herself and her ways of understanding her relation to the world." This definition, while useful for conceptualizing the term, focuses specifically on the mind as *the* locus of self-awareness. In privileging consciousness over corporeality, the body is left to the wayside as having little influence over our experiences. This split between mind and body has caused many feminists to wonder about the role of corporeality in our understandings of self, other, and society. Elizabeth Grosz (1994) challenges the mind/body split by affirming that the body and mind are neither two distinct entities operating in mutual exclusion nor entirely the same entity. Instead, they lie somewhere in between, as the metaphor of the Möbius strip helps to explain.

> The Möbius strip has the advantage of showing the inflection of mind into body and body into mind, the ways in which, through a kind of twisting or inversion, one side becomes another. This model . . . provides a way of problematizing and rethinking the relations between the inside and the outside of the subject, its psychical interior and its corporeal exterior, by showing not their fundamental identity or reducibility but the torsion of the one into the other, the passage, vector, or uncontrollable drift of the inside into the outside and the outside into the inside. (xii)

As Grosz explains, this torsion of mind and body is what produces embodied subjectivity, where psychical interior and corporeal exterior connect to and wind around each other, deepening an understanding of the self in relation to the social world. Lois McNay (1999, 98) offers a similar approach to subjectivity, explaining that the body is "the point of overlap between the physical, the symbolic and the sociological, the body is a dynamic, multiple, mutable frontier. The body is the threshold through which the subject's lived experience of the world is incorporated and realized." The concept of embodiment thus enables feminists to talk about the body without reducing women or girls *to* their bodies, avoiding charges of biological determinism that have plagued feminist discussions of the corporeal self for decades.

The feminist concept of embodiment is an intriguing one in relation to style. After all, the body is never naked in Western society; it is always dressed. As compulsorily dressed bodies, where does style fit into embodied subjectivity? Style becomes the observable, public layer that covers the hidden, private layer of the corporeal self. But like mind and body, these two layers are neither wholly distinct nor wholly unitary. Body and style complement each other, reinforce each other, and socially sculpt each other. While both are constructed out of different materials, they, too, drift uncontrollably together, integrating the interior layer of the body with the exterior layer of clothing that offers it protection and concealment. Style is the public face of the body, acting as the body in social space. Style is the body's emissary and messenger, making the body visible, mappable, and readable. Because of these torsions, Joanne Entwistle (2000, 10) suggests that dress, the body, and the self are so closely aligned that they "are not perceived separately but simultaneously." When we enter a room, we are seen not as bodies, but as dressed bodies. Clothing is that which reveals and releases the physical frame, cloaking and confining it, shaping it and giving it definitions, margins, and creases. Clothing is what directs people's attention to the body and what causes the body to become perceptible and intelligible. Clothing is also what enables others to evaluate us as bodies. In our clothes, we are judged, looked at, wondered about, envied, remembered, discriminated against, lusted after, admired, respected, and ignored. In style, we are recognized and misrecognized, understood and misunderstood, made familiar and rendered exotic and strange.

The opinions we register about the dressed bodies of others hovers just below the level of conscious thought. But this judgment works both ways; before we open our mouths, style enables others to glean information about us (Lurie 1981), including, perhaps, our cultural background, our social class, our age, and our sexuality. While we may use style to conceal or confuse these things, we also use it to purposefully signal lifestyle affiliations, such as musical taste, subcultural connections, and our level of commitment to commodities, consumerism, and stylishness itself. As Pierre Bourdieu (1984) explains, such nonverbal cues are a part of our *habitus,* our unconsciously learned embodied dispositions. For McNay (2000, 35), habitus is the internalization of social structure or the "living through" of one's social circumstances in a way that is discernible to others.[8] Through habitus, we imagine that we "know" someone because we like their style, the "cut of their jib," or their personal taste, making them seem like-minded and familiar.

Given these articulations, style functions as a powerful form of embodied subjectivity that acts as social skin in social space. As social skin, style is a membrane of permeability that enables us to transfer something of ourselves into the social world, while simultaneously enabling the social world to transfer something of itself into us. Style encompasses constructions of gender, race, ethnicity, class, and sexuality, constituting us in particular ways that limit or increase our social visibility. As we shall see in subsequent chapters, certain girls could "get away with" just about anything in relation to style, while others struggled to find their niche or a threshold of acceptability that would grant them access to particular subject positions. But style also enables something of us to filter into the social world, where we can make ourselves known as particular kinds of people who are positioned in particular ways. As a malleable and accessible form of embodied subjectivity, style enabled girls at ESH to creatively carve out subject positions for themselves, sometimes enabling them to successfully transform from one subject position to another, or to use various and contradictory subject positions to play off against each other in an effort to keep people guessing as to who they "really" were.

As social skin, style acts as a conduit between the self and the social world, public and private life, fantasy and reality, desire and action, visibility and invisibility, and interiority and exteriority. It is one of the key ways that meaning is granted to the body, fashioning the corporeal self into a fluid social text that we hope says something about who we are and how we want to be seen and treated by others. In this way, style is both an entrance into and an exit out of our shifting and contextual identities. But beyond these complex social processes, style is, as Kaja Silverman (1994, 191) suggests, quite simply a "necessary condition of subjectivity." Style announces to others what "kind" of relationship we are going to have with them and what kind of relationship we (hope to) have with the world. Our social visibility is wrapped up in style; our identities are contingent upon it.

FRUITFUL THEORETICAL CROSSROADS

As a study of how girls use style to negotiate their identities in the school, this book is situated at the fruitful theoretical crossroads of girls' studies and feminist sociology of education. Gayle Wald (1998, 587) describes girls' studies as a "subgenre of recent academic feminist scholarship that constructs girlhood as a separate, exceptional, and/or pivotal phase in female identity formation." Drawing on cultural studies and its interdisciplinary theoretical perspectives, girls' studies is concerned with how girls take up the texts of girls' culture in order to construct and negotiate diverse and alternative subjectivities.[9] As Alison Jones (1993, 159) suggests, if "girls become 'girls' by participating within those available sets of social meanings and practices—discourses—which define them as girls," then the project of girls' studies is to recognize that there "is no one way in which girls as a group, or as individuals, can be fixed in our understanding." In exploring girlhood as a social and cultural construction, the field of girls' studies goes beyond developmental models by asking, "how and why are girls what they appear to be at a particular moment in a given society" (de Ras and Lunenberg 1993, 1)? Girls' studies, to paraphrase Valerie Walkerdine (1990), focuses not on reifying a unified notion of girlhood, but on dismantling the fictions around which the category "girl" has crystallized.

The field of girls' studies has tended to focus on analyses of popular culture as a way of exploring girls' resistance to "emphasized femininity" (Connell 1987),[10] including subcultural and alternative forms of girlhood.[11] These forays into girls' engagement with cultural texts have been influenced by cultural studies, a field that draws from a diverse range of theories and methodologies. But as Angela McRobbie (1996) notes, cultural studies has been criticized for being overly concerned with texts. From a sociological perspective, cultural studies can seem indifferent to the ways in which social worlds are organized and how meaning making is made possible through such organizations. Cultural practices are often analyzed independently of the people, places, and social formations that make them possible and meaningful in everyday life.

Conversely, cultural studies theorists have critiqued sociology. Focused on ethnographic research that often seeks access to "true" voices, sociology has been accused of ignoring the discursive formation of the subject and the crisis of representation brought on by the "linguistic turn" in the social sciences. Sociology is marked by investigations of group interactions within institutional organizations, such as the school and the family, as well as qualitative methodologies that seek to paint an "honest" portrait of social life by presenting the voices of those in it as "transparently meaningful" (McRobbie 1996, 32). The difference between cultural studies and sociology is centered on the difference between the "cultural" and the "social," where the former "refers to objects and practices which engage girls in meaning-making" and the latter refers to the materiality of girls' lives and their "lived" realities (Currie 1999, 284).

Cross-pollination within these disciplines over the last two decades has caused Michèle Barrett (1992, 205) to observe a "turn to culture" in feminism, based on a move from sociological "things" to the "words" of cultural texts. For Barrett, the kind of feminist sociology that has been gaining attention has "shifted away from a determinist model of 'social structure' (be it capitalism, or patriarchy, or a gender segmented labor market or whatever) and deals with the questions of culture, sexuality or political agency." Feminist sociology of education has the potential to push this cross-pollination further by linking these issues to the institution of the school, where dominant notions of gender, race, ethnicity, class, sexuality, age, and nationhood are reified in power relations produced through curricula, classroom practices, and bodily discipline. Feminist sociologists of education, particularly critical feminists who work toward social change,[12] emphasize the school as a primary site of unequal relations of power that are played out in gendered ways.

But for the most part, feminist sociology of education does not link the "pedagogical regimes" (Luke 1996, 8) that construct girlhood in the school to the cultural practices that are central to girls' culture.[13] Typically, such analyses stop short of analyzing the meaning-making practices of girlhood and instead focus on the gendered pedagogy of the classroom or the gendered structure of the school as an institution. However, as Carmen Luke (1996, 1) points out, pedagogy cannot be contained as an "isolated intersubjective event." Luke expands the traditional definition of pedagogy as teaching and learning to mean "the many pedagogical dimensions of everyday life implicated in the constructions of gendered differences and identities" (4). This expansive view pushes beyond school curricula and reaches into hallways, bathrooms, parking lots, and smoking areas—social spaces that are solidified by girls' engagement with cultural texts.

There is a disconnect between sociological analyses of gender in the school and the textual investigations of girls' studies. Situating this book at the fruitful crossroads of girls' studies and feminist sociology of education thus facilitates a sociological and ethnographic understanding of girlhood as it is constructed within the institution of the school, while at the same time engaging with the texts of girls' culture as they are used by girls to negotiate that construction. Working at this theoretical crossroads enables

me to explore the cultural texts that girls take up as they perform and negotiate their identities, while focusing on the school as a constituting institution that positions girls in specific (gendered, raced, classed) ways.

As a product of discursive processes, girlhood is slotted into models of intelligibility that render it the "passive object and neutral product
of larger and more important historical changes and voices" (Driscoll
2002, 4). With Driscoll's words in mind, I know that I am part of the
process that produces and perpetuates knowledge on girls. As Mary
Celeste Kearney (1998, 306) warns, researchers "contribute to the established story of a girl's life." But the aim of this book is not to prescribe
how girls should be or what girls should do. Rather, the aim is to explore
how girls are positioned and position themselves within the school,
and how these positionings necessitate an engagement with the everyday
cultural practices of girls. More specifically, my aim is to challenge the
single certainty surrounding girls by highlighting the messy and multiple versions of identity that they experience through style. With this goal
in mind, it is necessary to explore how I co- and re-created the stories
that appear in this book. Ethnography was the methodology that took
me into a world that, however familiar, was indeed still very strange. As
much as I may have wanted to participate in ESH's social world (if only
for a fleeting moment), there was never any doubt in my mind, or in the
minds of girls, that I was—and remained—an outsider.

WHAT ARE YOU DOING HERE, ANYWAY?

On my eleventh day at ESH, Geraldine, the curt but efficient secretary
in the office, told me that I no longer needed to pick up a visitor's badge
when I signed in each morning. "People are getting used to you," she
said without looking up. At first, I was pleased to have been "gotten used
to" and to have somewhat mitigated my role as the "anthropological
stranger" (Agar 1980). But as I left the office without the badge, I realized that I had come to rely on its easily understood designation in order
to negotiate my own identity in the school. When the badge fell from my
shirt one day, Casey called out to me from her lunchtime spot on the
floor. "Shauna, you dropped this!" Walking back, I thankfully scooped
it out of her hands and replied only half jokingly, "without this, I'm
nothing!" With the badge, I had a clearly defined purpose—one that
was provisional and temporary, yet comprehensible and official. I was a
visitor. But without the badge, it became increasingly difficult to explain
to students why I was still hanging around their school.

As an ethnographer, I fell through the cracks of familiar social categories. There was no badge that I could have worn to adequately sum up

my attendance at ESH. But occupying a liminal space "in-between the designations of identity" (Bhabha 1994, 37) meant that I was easily slotted into just about any category in attempts to make sense of my presence in the school. From time to time, I was positioned as a teacher—asked to explain things to the class, spell hard words, roam around and offer assistance, and "cover" for a while. I was also positioned as a student by those same teachers who sometimes saw me as one of the many bodies before them in class, where I sat day after day, taking notes, putting up my hand on occasion, and worrying that I might get into trouble for whispering to the girl beside me. Even students sometimes saw me as one of them. As Steven, my dishwashing partner in ninth grade Foods and Nutrition, noted, "You're like a kid. We're used to you, but you don't have to do the work."

I was also taken for a reporter or a writer. During one particular lunch, I was hanging out in the drama room, a refuge for some of the self-defined "alternative" students in the school. Someone had put on a tape of Arlo Guthrie's *The Motorcycle Song*. About fifteen of us were lounging on chairs and grungy couches in the open space of the studio. I had come to know most of the students in the room, but one, Davis, was new to me. He was a skater with curly blonde hair that easily escaped his black toque. Between moments of singing, laughing, and the general mayhem that filled the room, he hesitantly turned to ask me if I was a "student or a teacher," not really sure if either category was suitable. Someone called out, "she's writing a book!" Interested and a little defensive, Davis asked, "What, like, are you looking for the *stereotypical* teenager?" I explained to him that I was a university researcher interested in the social world of ESH and girls' identities in the school. Still a little puzzled, Davis decided to simply tease me, asking, "Is it going to be like *Degrassi Junior High?*"[14]

More commonly, I was positioned as a girl/friend. My own style in the school was different from that of most girls, but not sufficiently different enough to draw too much attention. I wore jeans and skirts, T-shirts and tank tops in warmer weather, sweaters and blouses in colder weather, and a favorite pair of chunky black platform boots all year round. At thirty-two, I was youthful looking and my genuine interest in girls' culture helped to legitimate my conversations with girls. On Wednesday mornings, girls would seek me out for the obligatory *Buffy the Vampire Slayer* rehash, on Thursdays it was the *Gilmore Girls,* on Mondays it was their tales from the weekend: drinking, dancing, clubbing, dating. But this "insider" knowledge did not always make me popular and every now and then I found myself being tested to see if my loyalties were intact. I was sometimes expected to go along with a lie in order to get students out of a jam or to cut class and head to the mall at a moment's notice. Girls striving for an even deeper show of solidarity would ask me to "party" with them on the weekends or buy them alcohol at the local liquor store. My polite refusals were often taken in stride as most girls understood that I was an adult, even if they did not comprehend what it meant to do ethically bound, university sanctioned research.

Freddie, a senior and admitted slacker, once asked, "What are you doing in *my* school?!" More commonly, the question was issued to me in this form: "What are you doing here, anyway?" Often asked out of curiosity, it was certainly a good question and one that I had to contend with on a series of imbricated levels. What was I doing in the school, in theory? This question necessitates an engagement with the theoretical and methodological choices I made prior to entering ESH and how those choices framed the study. On another level, I wondered what I was doing here, in the field. In other words, how did my hoped-for theoretical and methodological framework fare upon entering the social world of the school and becoming a part of ESH culture, however partially and temporarily? This question necessitates an engagement with how girls' innocent inquiries, curious comments, and thoughtful insights pushed me beyond my own conceptions of the project to create something that, like the subjectivities of those involved, continuously shifted and diversified. Finally, what am I doing here, in the ethnographic text? This question necessitates an engagement with the epistemological desire to produce knowledge about girls and to become a part of the discourse that constructs how girls are perceived. In the remainder of this chapter, I explore these three imbricated strands of ethnography—theory, field, and text—in order to make plain the decisions, indecisions, and revisions that have shaped this book.

DOING FEMINIST POSTSTRUCTURAL ETHNOGRAPHY

I came to ESH to conduct a feminist poststructural ethnography on girls' negotiations of identity through style. I felt that a feminist poststructural ethnography was well suited to a deconstructive text, one that focused on keeping the signifier "girl" in play, rather than constructing further binaries, such as "good" girls/"bad" girls, "nice" girls/"mean" girls, and "normal" girls/"abnormal" girls. As much as possible, I wanted to avoid this prescriptive tone. A feminist poststructural stance seeks to engage in Derrida's (1978, 293) challenge to wriggle beyond the closure of binary thinking, toward the "as yet unnamable which begins to proclaim itself" when categories such as "girl" and "girlhood" remain undecided.

This kind of ethnography is not concerned with smoothing out the edges in order to generate seamless research, a practice that produces "homogeneity, coherence, and timelessness" (Gonick 2003, 30). Instead, it resists closure by reflexively analyzing and maintaining the unevenness of the data.

Ethnography, as James Clifford (1986, 2) explains, has its roots in the "persistence of an ideology claiming transparency of representation, immediacy of experience." Its initial premise was to portray the "Other" in a way that made them knowable, digestible, and safe (Fine 1994a, 1994b; Said, 1978; Smith 1999). Such an ethnography, generally characterized by the terms "positivism" and "naturalism,"[15] claims to produce a rendering of culture that is unmediated. As Trinh Minh-ha (1989, 53) notes, it relies on "a neutralized language that strips off all its singularity to become nature's exact, unmisted reflection." The ethnographer retains his or her guise as the "anthropological stranger" and remains invisible both in data collection and in the write-up of the ethnographic text. This invisibility is meant to suggest a suspension of judgment, value-neutrality, and objectivity, as the ethnographer has not jeopardized the study by becoming too personally involved. Donna Haraway (1988, 584) calls this view the "God trick," offering "a vision that is from everywhere and nowhere, equally and fully."

The kind of ethnography described above has shifted significantly in the wake of critical, feminist, and anti-foundational influences. As an anti-foundational stance, poststructuralism offers a way to think through (though certainly not resolve) the dilemmas of the "God trick." Heralded by the "linguistic turn" in the social sciences, poststructuralism "focuses on the power of language to organize our thought and experience" (Lather 1991, 111). Language does not reflect reality by capturing an already existing social world in words. Instead, reality is produced by language and its ability to structure whole systems of thought. As Derrida (1973) suggests, language cannot point to any one reality as it can never point to one stable form of meaning. Rather, it points to a continuous string of possible and shifting meanings, deferring full presence. *Différance* is the name Derrida gives to this deferral, a concept that separates words, or signifiers, from their arbitrary designations, or the things they signify. The shift from language as it reflects reality to a reality that is produced through language signals the linguistic turn and the related crises of representation and epistemology in the social sciences. If meaning shifts depending on social context, then representation can only ever be a "temporary retrospective fixing" of contextualized truths (Weedon 1987, 25).

For Foucault (1972, 1977, 1978, 1980), language signifies socially constructed systems of knowledge. As I noted earlier, these systems, or discourses, produce social reality by enabling certain possibilities to exist, while making others unfeasible, illegal, or immoral. Discourse comes to organize how we think about ourselves and the world, creating forms of regulation that feel "normal" instead of the product of social, cultural, and historical constraints. Poststructuralism, as Deborah Britzman (2000, 30) notes, raises "critical concerns about what it is that structures meaning, practices, and bodies, about why certain practices become intelligible, valorized, or deemed as traditions, while other practices become discounted, impossible, or unimaginable." To do poststructural research is to foreground the impossibility of unmediated representation by reflexively analyzing the discursive forces in which researcher, researched, and research process are entwined. For Laurel Richardson (2003, 509), poststructuralism thus highlights "the continual cocreation of Self and social science."

Feminist thinking has pushed poststructuralism in new directions, working to infuse these deconstructive aspects with a critique of gender as it intersects with race, ethnicity, class, and sexuality. Feminist poststructuralism seeks to "question the location of social meaning in fixed signs" (Weedon 1987, 25), specifically as they pertain to the fixed signs that construct gender. As Davies (2000, 179) suggests, such analyses focus "on the possibilities opened up when dominant language practices are made visible and revisable." The categories "girl" and "woman" are kept in play not only to highlight their arbitrary designations within society but also to explore all the possibilities that have been foreclosed upon or left unexplored as a result of such constructions. Just as meaning is rendered unstable through *différance,* so too are gender and gendered categories that seek to fix girls and women in rigid modes of being. Feminist poststructural analyses work to crack open the limitations of gender by pointing to the discourses that enable these limitations to exist and thrive.[16]

Given my interest in moving away from the single certainties that surrounded girls in the press and in the public imaginary, conducting a feminist poststructural ethnography made perfect sense to me as I prepared to enter the school. But enacting it in the field was another issue altogether.

ALTERNATIVE CHOICES AND ACTS OF OMISSION

My desire to conduct a feminist poststructural ethnography meant that I needed to think of my methodology as more than just a passive plan of action, but as a politics in and of itself (Kelly 2000). I sought to challenge the imbalance of power that existed between girls and the larger mechanisms of knowledge production that forged single certainties around girlhood. A number of girls told me that they were glad I wanted to hear their opinions, as the media were always making them "look bad." During one discussion, Chrissie told me she wanted to be in the study because she was "sick and tired" of hearing negative things about girls in the news and how girls were just "open wallets" that marketers could take advantage of. Similarly, during a discussion with Abby, she told me that "marketers think girls will buy anything," which deeply irritated her.[17]

In an effort to conduct the kind of feminist research that I felt would challenge dominant discourses on girlhood, I had initially conceived of my study differently. Though I had always intended to study girls' negotiations of identity through style, I had originally thought that I would focus on "alternative" girls. I defined alternative girls as those who dressed against the "mainstream" or in ways that were not considered to be "fashionable" or "trendy." Alternative girls, I reasoned, could offer different models for viewing girlhood other than those represented by the press. But significantly, I had trouble finding participants who fit the bill. Teachers did not understand what I meant by "alternative," telling me that all of the girls in their classes were "really regular" and "just average." And in my initial interviews, I spent more time explaining what *I* thought "alternative" meant rather than focusing on how girls saw themselves in relation to the term. The typical response was, "alternative to *what?*"

During my first few weeks at ESH, I could feel that something was not working, but after talking with Ratch, a self-defined punk, I understood that I had to shift my study in a new direction. I had interviewed Ratch as an "alternative" girl, but she questioned my criteria for this classification. Some girls, she told me, looked alternative on the outside, but were really mainstream on the inside. And some girls, she continued, looked pretty mainstream on the outside, but were actually quite alternative on the inside. Alternative girls, she deduced, were impossible to define. A few days later, Ratch leaned over to me in English class and whispered, "I've been thinking about your study." "Yeah?" I said out of the side of my mouth, not wanting to get in trouble for talking during the lesson. "You know, *everyone's* alternative in their own way." I nodded. "Punks and goths and hippies are so mainstream. You can just go to the store and buy it." But, she explained, "if you choose to look different, you have to be prepared for other people to look and make comments. So it's a lot harder to look different than to [actually] be different."

Though Ratch did not use these words, she was trying to explain to me that the "alternative" category was unstable as it pointed to many possible meanings within many shifting contexts. She was trying to tell me that I was fixing meaning, rather than attending to the ways in which meaning becomes unfixed through its endless deferral. She was also trying to tell me that the word *alternative* was inadequate for describing any "true" reality in her school. In short, she was trying to tell me that my study did not make sense.

As a result of this exchange, I pulled away from trying to find "something" and instead began to ask different questions, such as, how do girls from a variety of social locations use style to negotiate their identities *here?* After all, the point of poststructural research is to "increase the circumference of the visible" (Søndergaard 2002, 202). Though the category "alternative" did come up, it was girls themselves who used the term in relation to their own understandings of identity in the school. By reframing my study as exploratory rather than as a quest to find something in particular, I was no longer limited to one framing of girls, a framing that worked against my feminist poststructural desire to generate rather than foreclose upon possibilities. My new open-ended focus was thus a direct result of the negotiations that took place between myself and research participants like Ratch, who had no trouble telling me when they felt I had gotten it wrong.

Feminist poststructural ethnography highlights the contradictory elements inherent in any research project by showcasing the "hesitant voices of participants who kept refashioning their identities and investments as they were lived and rearranged in language" (Britzman 2000, 31). My own hesitations and false starts were also a key component to the ethnographic project. My initial plan was to position girls "in a single textual location" (Walkerdine 1990, 198), but this intent would not have been compatible with what I actually learned or with my feminist agenda to highlight the significance of style for girls in the school from their own perspective. By leaving such tensions in the text, however, ethnography can ask "questions that produce different knowledge and produce knowledge differently, thereby producing different ways of living in the world" (St. Pierre and Pillow 2000, 1).

For Kamala Visweswaran (1994, 48), "[a]cts of omission are as important to read as the acts of commission constructing the analysis." These silences, fumbles, and false starts all culminated in the generation of this ethnographic text. I struggled to balance my desire to change people's minds about girls with an understanding that I was not on some "university rescue mission in search of the voiceless" (69). I was, however, engaged in a complexly knit social world in which I, as the ethnographer, was both powerful and powerless. Girls enacted power by refusing to talk to me, by challenging my questions and sampling, by rejecting my presence as a white, middle-class, university-educated researcher, and by seeking constant clarification on what I was doing in their school. During perhaps my most humbling moment as an ethnographer, Ernie told me that I was the topic of his short-story assignment. "What's it about?" I asked, somewhat flattered. He replied, "It's about a university researcher who goes back to high school in a pathetic attempt to recapture her youth." I smiled at Ernie and (stupidly?) asked if he felt that that was what I was doing at ESH. He paused for a second and then replied, "No! I just added the pathetic part for dramatic effect!" But whether he had meant it or not, Ernie's short-story plot forced me to continuously question why I had come to ESH—and to question my desire to be a part of a social world to which I most certainly did not belong.

As Marnina Gonick (2003, 22) suggests, the very real effect that students had on me and that I had on them made it hard to balance "the epistemological foundations of knowledge production, research ethics, and the hoped-for critical accomplishments of a feminist ethnographic project." But nonetheless, in order to say something, *anything,* about girls, I knew that a research relationship had to be entered into. And while there was a great deal to think about in the ethnographic process, I "did not want the complexities of the [girls] lives to be reduced to my history" (Skeggs 1997, 34). Though I questioned my motives for going back to high school (a thought that made most of my thirty-something friends cringe),[18] I knew that I had come to ESH to do research on and with girls—and to do the kind of research that I hoped would *not* reproduce the very issues that had driven me into the field in the first place.

TOWARD A GENERATIVE UNDERSTANDING OF GIRLS

The stories in this book were co- and re-created by myself and the girls, in the locale and the particular historical moment in which this research was conducted. Yet in spite of this generativity, written culture has a way of trapping those who are represented within the text. The girls written about here have graduated and moved on to other things. They are not where I left them, in broken blue and orange plastic chairs, crammed into fogged-up, smelly bus shelters, and sitting cross-legged amidst the rotting apple cores and empty chip bags that littered the hallways of ESH during lunchtime. They have aged, grown, and changed styles. As I write this book, longer shirts, jeans with higher rises, and a retro 1980s look have made a comeback. A new economy of style is already in the works, as are new ways for girls to perform girlhood in the school. But it is my hope that this book, however historically, socially, and culturally specific, will generate a deeper understanding of girls' cultural practices.

The knowledge generated here is situated, but there is a drive toward generating an understanding of girls that is linked to generations of girls past, present, and future. This link occurs through the cumulative effects of discourses that have constructed, are constructing, and continue to construct girls, perpetually reaching backward and forward through time.

The stories of girls' negotiations of identity that follow are written in order to offer glimpses into girls' meaning-making practices, but also to generate new possibilities for girlhood within feminist qualitative research. I am hoping to generate more questions than answers, different responses from readers (who inevitably bring their own knowledge of girls to the text), a contextualized reading of girls' cultural practices, critical engagement of the power imbalances between girls and those who would define them (such as myself), and more ways of knowing and more things that can be known about girls. Finally, it is my hope that this ethnography will generate others, as other studies have most certainly generated it.

Girls engaged in generativity as they performed their identities in the social world of the school. They dressed for parts in which they were cast based on the subject positions they occupied. But within these seemingly fixed social roles, girls continued to negotiate the parts they wanted and yearned to play through style. Sometimes they were successful, making transitions effortlessly from one school identity to another; sometimes they were seen as "posers" and forced to return to an earlier incarnation of themselves; sometimes they occupied several contradictory identities at once; sometimes they forged hybrid identities, refusing to occupy just one at a time; and sometimes they fantasized about possibilities that they knew existed, but were afraid to make happen. As Sandra Weber and Claudia Mitchell (1995, 62) suggest, clothing is a complex amalgam of social needs and fears, signaling "a proclamation of resistance, a mode of innovation or becoming, a reconciliation, a desire to belong, or a surrender." Such articulations suggest that style is a crucial feature in girls' everyday lives.

In order to make connections between style and girls' negotiations of identity in the school, I observed, interviewed, and hung out with girls from a variety of social locations. I rotated through thirty-seven classes in order to meet girls from diverse backgrounds, and eventually settled on a semiregular schedule that offered me the widest "view" of the school. I focused on interviewing, observing, and hanging out with girls from three diverse programs: French Immersion, an academic program composed of mainly white students who were middle- and working-class; the "Regular" program, ESH's racially and ethnically diverse "main" student body, composed of middle-, working-class, and working-poor students; and girls in the Aesthetician program, otherwise known as "beauty school," a vocational stream that was composed of working-class and working-poor girls from racially and ethnically diverse backgrounds. This selection of programs offered me the opportunity to get to know girls from a variety of economic locations who were both academically and vocationally positioned within the school, as well as girls from a variety of racial and ethnic backgrounds. I formally interviewed twenty girls (many of them twice), but informally chatted with dozens more through the camaraderie that developed in the classroom day in and day out.

Whenever possible, I participated in the activities of the classes I attended, analyzing poetry, calling out suggestions for improv skits, taking part in language games, where I was pegged by Mr. Murphy as a member of the "Boggle generation" (a title which, sadly, I did not live up to), conducting science experiments, discussing Canadian history, and learning about "up-dos" and facials. I also spent time with girls at lunch, hanging out with them in their various corners of the school: stairwells, classrooms, the cafeteria, bus shelters, the smoking area, and dented rows of locker banks that lined every hallway. I attended assemblies, evening events, and chatted with students as we meandered home at the end of the day. I also happily took fashion advice from girls. Suffering from style paralysis each morning before heading off to school, I stood before my closet like a deer caught in headlights wondering how to achieve the look I was going for—though I never really understood *what* that look was exactly. How did I want to represent myself as an "ethnographer," as an "adult," as a "girl"? Though style paralysis often made me late for the crosstown bus that took me to ESH, I tried to follow Gianna's advice. Gianna was about to graduate from the Aesthetician Program and had a strong sense of her own style. She explained to me in no uncertain terms that it did not matter *what* I wore, but *how* I wore it. "Rock what you're wearing," she said. When I asked her what she meant, she responded with a snap of her fingers, "own it."

While this book focuses on the ways in which girls used style at one particular school in order to construct and negotiate their identities, my broader aim is to highlight the complexity of style in girls' lives by resignifying it as a significant cultural practice. To that end, chapter 2 focuses on how girls' style has been infused with certain meanings through academic, professional, and commonsensical discourses. In tracing how girls' style means as ideology, conformity, and pathology, I work to reconfigure style as a point of convergence for girls and power, where girls are both influenced and influencing in relation to global capitalism, popular culture, and professional pronouncements on what constitutes a "normal" and "healthy" girlhood. Turning to the school in chapter 3, I explore ESH as a multiply constituted site that was continuously constructed through internal and external reputations. In refusing to smooth out these rhetorical performances (including my own), I highlight not only how the school's identity is repeatedly formed, but also how the rhetorical performances of the school were inextricably linked to girls' own performances of identity.

Chapters 4 and 5 focus specifically on girls' negotiations of identity through style. Chapter 4 highlights the hesitant processes of subject positioning as girls used style to forge identifications between themselves and others. Here, I explore style as a pivotal feature in the formation of social groups, as well as the ambivalences girls experienced in their desire to be recognized as particular "kinds" of girls based on the social categories that certain styles seemed to guarantee. Chapter 5 focuses on style as an expression of agency. Girls spoke of their "images" and how they wanted others to see them, and used style to creatively negotiate their positionings within discourses of gender, race, ethnicity, class, and sexuality. In this chapter, I also explore girls' negotiations of an image as an expression of power, including the power to subtly negotiate cultural and religious boundaries, the power to instill fear, and the power girls said they experienced as sexual subjects. In chapter 6, I conclude by revisiting the single certainties surrounding girlhood in the press. While style is the focus of this book, I also explore other modes of reductionism that have culminated in a backlash against girls in the early twenty-first century. I suggest that a social science dedicated to uncertainty and the messiness of everyday life offers the possibility of seeing beyond single certainties, moving us closer to a generative understanding of girls and girlhood.

CHAPTER 2

GIRLS' STYLE AS A POINT OF CONVERGENCE

> Like, I'm sure if adults went out and talked to us, they'd go, "I guess
> girls are a lot smarter than we thought they were."
>
> Abby, "bad-ass-schoolgirl-business-woman" style

BEFORE MOVING ON TO A DISCUSSION of how girls used style to construct and negotiate their identities in the school, it is useful to place the
topic within a broader framework by asking an awkward question: *How*
does girls' style mean?[1] In asking this question, I am hoping to move
away from the customary question, *what* does girls' style mean? As
I noted in chapter 1, it is impossible to fix style within denotative values,
determining, once and for all, what items of clothing indicate in a stable,
pregiven form. Style must be read as a shifting system of signs that gains
meaning only as it becomes meaning*ful* within a given context. To suggest that explicit values or "maps of meaning" may be applied to this or
any cultural practice only severs it from girls' fluid and unpredictable
expressions of identity.

Instead of asking what girls' style means, then, I have chosen to ask a
different kind of question in order to highlight girls' style as a discursive
construction that has been infused with meaning through a variety of
academic, professional, and commonsensical discourses. In asking how
girls' style means, I aim to unearth dominant understandings of girls'
style and how those understandings have come into existence. How has
girls' style come to mean a set of truths that claims something about who
and what girls are? And what do these meanings have to say about the
way our society looks at, treats, and values girls?

In asking how girls' style means, I am also working toward its resignification. While this book focuses on the ways in which girls engaged in

the practice of style at one particular school, my broader goal is to highlight the complexity of style by resignifying it as a cultural practice that
has meaning in the lives of girls. By implication, I am suggesting that
girls' style has been and continues to be devalued. This chapter traces
how this view came into being and questions why such a central feature
of girls' culture has been so readily and easily dismissed. After all, style is
neither a marginal cultural practice, nor one engaged in by only a select
group of girls. All girls practice style, even if they conceive of their
styles as minimal, barely there, or under the radar. Yet when I explained
the study to people at academic conferences, dinner parties, and social
events, I learned to feel a little uneasy about their possible responses.
"But it's *just* style" was a common refrain. Reading between the lines, I
felt that people were trying to tell me that I should be studying something more serious, something with more substance and depth. Sometimes, people would offer up possible topics that I might rather investigate,
such as the "mean girls" crisis, bullying, or girls' "plummeting" self-esteem
in the teenage years.

Others took a more argumentative tone, suggesting that studying
girls' style was an exercise in futility. "Girls won't be able to tell you anything about it. They're clueless." Or similarly, "Here's everything you
need to know about girls and style . . . they dress the way marketers tell
them to dress." And others still would let me know in no uncertain
terms that I was misguided in my desire to construe style as important,
given how "harmful" it was for girls to be dressing in ways that encour-
aged the "wrong" kind of attention. In such conversations, I heard real
anger and frustration directed toward girls for their perceived naiveté
and lack of awareness about what they were buying and how they were
choosing to represent themselves in public. As one person casually noted,
"I'm not sure what there is to learn about a bunch of girls who hang out
at the mall all day."

What surprised me most about these reactions was the bleak view
some people had of girls' cultural practices, particularly one so crucial
to girls' identities. It seemed to be an all-or-nothing proposition: either
girls were doing something good or they were doing something bad;
either girls were doing something positive or they were doing something
negative; either girls were powerful or they were powerless. But such
dichotomies do not make sense given the intricacy of girls' engagement
with cultural forms. Instead, I argued for a more nuanced and complex
understanding of girls' engagement with cultural practices. While I readily
acknowledged that marketing, media, and global capitalism influenced
girls, I questioned a view of girls' experiences that so consistently and
resolutely condemned them as negative. I wondered how this impression

of girls gained such a hold in our society and why is it that so many people feel justified in disparaging the things that girls do. Certainly, men's, boys', and to some extent women's cultural practices are not given the same disapproving treatment.

As Driscoll (1999, 177) aptly notes in relation to the Spice Girls' phenomenon of the 1990s, anything "perceived as a girl thing" is easily dismissed as "delusion," "even compared to bigger market conformities . . . like wearing Nike or drinking Coke." Time and again we see this dismissal in action as girls' culture is either ignored or criticized. Walkerdine (1993, 15) highlights these sentiments by noting that girls are always, "accused of being too mature, too sexual, too hard-working, too rule-following, not rational enough, always in trouble—always too something and not something enough." Similarly, Christine Griffin (2004, 42) suggests that dominant meanings of girlhood function through a series of contradictions, making "the girl" an "impossible subject." She notes that girls are often represented "as having (or being) too little or too much; as too fat or too thin, too clever or too stupid, too free or too restricted." Meanings of girls' style operate within a similar set of restrictions and excesses. Generally speaking, girls' style is perceived to be too frivolous, too mainstream, too co-opted, too conventional, too sexy, too inappropriate, and too boring. It is also seen as not political enough, not resistant enough, not agentic enough, not powerful enough, and not "authentic" enough to be taken seriously by academics. These common perceptions are a product of how girls' style means. For the most part, girls' style has come to have three dominant meanings in North American society. It means as a form of ideology, or false consciousness; it means as a form of conformity, or submissive obedience; and it means as a form of pathology, or deviance symptomatic of psychological trouble.

As a form of ideology, style is viewed as something that girls do unconsciously and without critical thought. Through a Marxist lens, style represents false consciousness, where girls allow themselves to be taken advantage of by marketers and the global capitalist empire. Here, girls are constructed as "cultural dupes" (Hall 1981) who are easily persuaded to shop till they drop, buying anything and everything that is marketed to them. As a result, consumerism is merely a programmed response, rather than girls' engagement with the raw materials of identity.

As a form of conformity, style is how girls emulate others. Girls are viewed as boring and passive in a one-to-one equation that suggests they simply do what they see. Here, girls copy favorite pop and film stars, palely echo boys' subcultural resistance, and remain deeply rooted in mass cultural submissiveness, where they are constructed as "media sheep" who follow rather than lead. As a result, girls' style is constructed as "inauthentic," without punch or politics, instead of a creative form of self-expression that has the potential, however subtly, to make an impact within a particular context.

And as a form of pathology, style is viewed as a symptom of girls' psychological distress. Sartorial choices that stand outside of conventional femininity, such as masculine looks, subcultural affiliations, piercings, tattoos, unusual hair colors, and overtly "sexual" styles are cause for alarm and even therapy. Here, style represents both ideology and conformity, but adds the idea of girls' mental and physical health. Girls are seen to be easily swept along by media and peer pressure, but this influence is constructed as dangerous to girls' well being. As a form of pathology, style is symbolic of girls' desperate need for attention from boys and frantic need to fit in with other girls. Style thus positions girls as victims, not just of marketing and media, but also of girlhood itself. Through these dominant meanings, girls' style has been drained of substance, becoming an empty signifier that denies girls' uses of style as a cultural practice infused with power. Instead, girls are always power*less:* powerless to understand what is going on, powerless to think for themselves, powerless to take action in the hegemonic sphere of popular culture, and powerless to push the boundaries of "acceptable" femininity without being viewed as abnormal. Given these meanings, girls' style has been and continues to be constructed as nothing more than a form of cultural domination that enacts power *over* girls, without any possibility for the practice of power *by* girls.

For Foucault (1977, 1978, 1980), such a vision does not do justice to the ways in which subjects are always enmeshed in power relations. Instead of a top-down approach where subjects are merely acted upon by structural forces, for Foucault (1978, 93), power is "produced from one moment to the next at every point, or rather in every relation from one point to another. Power is everywhere; not because it embraces everything, but because it comes from everywhere." If power is everywhere, then it can never be strictly the domain of institutions, individuals, or groups. As a set of relations, power is neither static nor linear, neither a commodity nor a possession. Instead, power "is employed and exercised through a net-like organisation" and we "circulate between its threads" (Foucault 1980, 98). Power thus flows multidirectionally, carried in discourses that establish truths and construct knowledge. As Foucault (1978, 101) explains, discourse is the thing that "transmits and produces power; it reinforces it, but also undermines and exposes it, renders it fragile and makes it possible to thwart it." Power, as a permanent feature of discourse, is thus equally thwartable, making it more than just a tool of coercion and subjugation, but a fluid and relational process.

Such a definition of power means that girls can never be merely powerless in their engagement with cultural forms. Girls circulate between power's threads, not as pawns, but as participants in power relations. While girls are subjected to constituting forces, such as marketing, media, global capitalism, popular culture, and professional definitions of "who" and "what" they can be, they are also subjects who have been enacted through power, able to navigate and shape these forces—to be both influenced and influencing at the same time. As Elizabeth St. Pierre (2000, 486) suggests, "though discourse is productive and works in a very material way through social institutions to construct realities that control both the actions and bodies of people, it can be contested." Girls are actively engaged in power relations that enable them to react and challenge, not simply receive and digest. For Foucault (1978, 95), contestation always takes a multitude of forms, as power relations are contingent upon "a multiplicity of points of resistance: these play the role of adversary, target, support, or handle in power relations." Power and resistance thus continuously produce each other, as the flow of power depends on contestation for its continued dispersal.

Given this formulation, the cultural practice of style is here viewed as a threshold of contestation for girls. Rather than something that is done "to" girls, style, as the word *practice* suggests, is something that girls *do*. This "doing" configures style as a "complex strategical situation" (Foucault 1978, 93) that moves it beyond prevailing meanings of cultural domination. As I noted in chapter 1, style operates as social skin, or a membrane of permeability that enables girls' subjectivities to filter into the social world, while simultaneously enabling discourse to filter into girls. This constant and steady flow of power infuses style with unique qualities. While there are certainly other modes by which power operates in girls' lives, style is distinct in that it is constantly observable and revisable, constantly under surveillance and up for discussion, constantly in the press and on the minds of adults, constantly a part of girls' public and private lives, constantly at the hub of consumer girls' culture, constantly a focal point of girls' social organization, and constantly acting as social skin in social space. It is here, at the point of convergence, where style ceases to be a symbol of girls' subjugation, and instead becomes an expression of power in girls' lives.

There is little question that girls are deeply affected by sociocultural forces that seek to interpellate them as "good" consumers, "good" citizens, and "good" girls. We are all, to greater or lesser extents, enmeshed in global capital's empire, marketing manipulation, media influence, and professional pronouncements on what constitutes a "normal" existence. But in tracing how particular views of girls' style have come into view, it is my hope to contribute to its resignification as an important feature in girls' lives, not just as a form of fun (though certainly that), but also as an important cultural practice that acts as a point of convergence between girls and power in our society. In exploring girls' style as ideology, conformity, and pathology, I aim to show how girls are "produced as a nexus of subjectivities in relations of power which are constantly shifting, rendering them at one moment powerful and at another powerless" (Walkerdine 1990, 3).

In tracing how girls have been constructed as powerless in relation to style, I also discuss how girls are powerful: as consumers, as a market, as participants in the hegemonic sphere of popular culture, as creative infusers and resisters of trends, and as embodied subjects pushing the boundaries of conventional femininity, where girlhood itself is being resignified. These forms of power do not negate the influences of sociocultural forces in girls' lives—they merely complicate them, suggesting that there is more to girls' style than dominant meanings imply. As a discursive analysis of how girls' style means, this chapter is about the ways in which "knowledge, truth, and subjects are produced in language and cultural practices as well as how they might be reconfigured" (St. Pierre 2000, 486). In reconfiguring style as a point of convergence for girls and power it is my hope to contribute to a deeper understanding of girls' cultural practices generally, and to expand current thinking on the importance of "girl things."

"CLUELESS": GIRLS' STYLE AS IDEOLOGY

In the 1995 film, *Clueless,* Cher Horowitz (played by Alicia Silverstone), is the quintessential shopaholic. Written and directed by Amy Heckerling, the film opens with a montage that focuses on Cher's group of friends hanging out, laughing, and having fun. One of the key expressions of this camaraderie is shopping at the mall. Through a voice-over, Cher addresses the audience. "So OK, you're probably thinking, 'Is this, like a Noxzema commercial, or what?!' But seriously, I actually have a way normal life for a teenage girl. I mean I get up, I brush my teeth, and I pick out my school clothes." The soundtrack shifts to David Bowie's, *Fashion Girl,* as we see Cher in her bedroom selecting an outfit for the day by inputting possible choices into a computer. Once she has achieved a good "match" between a top and a bottom, she scrolls through her battery-operated closet until she finds what she is looking for. As the audience sees her massive collection of clothing whizzing by on a conveyor belt, we immediately understand its significance in Cher's life. Style is the first and most important consideration of the day. Style is everything.

While the film shows girls to be engaged in a variety of activities, including matchmaking, partying, exercising, dancing, and schoolwork, its defining feature—and the source of its most amusing moments—is its correlation between girls' culture and consumerism. In the film, these two things do not just exist in a symbiotic relationship; they are construed as one and the same thing. In a smart script that mocks the typical representation of teenage femininity, consumerism is represented as the very lifeblood of girlhood. Cher and her friends live to shop; it brings them joy, comfort, and inner peace. They shop to bond with each other, they shop to get over boys, and they shop to clear their heads. They also shop as a good deed. When Cher and her best friend, Dionne, want to help the new girl, Tai, fit in with the popular crowd, they take her shopping. When Tai protests, Dionne responds, "Oh, c'mon! Let us! Cher's main thrill in life is a make-over, okay, it gives her a sense of control in a world full of chaos." Shopping is thus configured as a worldview in the film, becoming not just *a* form of girls' culture, but girls' culture itself.

Through acts of consumption,[2] the script cleverly plays with the construction of girlhood as a "clueless" state of being. When a fight she has with Tai devastates Cher, she wanders the store-lined streets of Beverly Hills, looking for solace. In this voice-over, we see Cher struggling to deal with her problems, but losing focus when a cute top suddenly appears in a store window.

> Everything I think and everything I do is wrong. I was wrong about Elton, I was wrong about Christian, and now Josh hated me. It all boiled down to one inevitable conclusion, I was just totally clueless. Oh, and this Josh and Tai thing was wigging me more than anything. I mean, what was my problem? Tai is my pal, I don't begrudge her a boyfriend, I really—[sees top in a store window] *Ooh, I wonder if they have that in my size.*

In the film's key moment of recognition, Cher is still shopping. While she is about to have an epiphany that will ameliorate much of her cluelessness, the revelation is easily brushed aside—if only for a moment—to wonder if a particular item of apparel will fit her. Her lack of awareness is thus perpetuated by her desire for clothing, her love of shopping, and her belief in consumerism as a way of life.

The film plays with girlhood as a social category defined by a lack of awareness and a detachment from reality. Cher putters along, doing her thing, shopping and enjoying herself, while the "real" world unfolds around her. Cher's inability to see what is going on is marked by her teenage girl lifestyle, which revolves around a consumerism devoted to clothing and beauty. Through the film, we understand style to be a form of ideology that keeps Cher and her friends from seeing the "truth" about their lives, a blindness shown to be the direct result of girls' naturalized position as clueless consumers. This connection is exemplified in one of the film's funniest moments. Shocked by the sexiness of his daughter's dress, Mel gruffly asks, "What the hell is *that?*" Cher, taken aback, replies, "A dress!" Unconvinced, Mel asks, "Who says?!" To which Cher sheepishly responds, "Calvin Klein." As this exchange humorously shows, for Cher, fashion is the highest truth. A dress is a dress because Calvin Klein says so. All other realities are secondary. As a result of the "linked acceptance of consumerism and femininity" (Schrum 2004, 24), style symbolizes girls' social, cultural, and political obliviousness and is thus constructed as a form of ideology in which girls are hopelessly enmeshed.

To say that girls' style is a form of ideology is to say that style is often viewed as a symbol of girls' perceived unconsciousness, particularly in relation to the capitalist system. For Marx, ideology is the very basis of capitalism, as it enables the ruling class to conceal "reality" from the working-class. This concealment is not an outwardly visible form of deceit, but functions beneath the level of everyday consciousness to hold the economic structure together. As Hall (1977) notes, it is exactly the "naturalness" of ideology that allows it to mean as commonsense, or as an invisible set of justifications for life as it is lived. Commonsense is often framed by terms, such as "that's just how it is" or "always was, always will be," suggesting the inevitability and inescapability of a particular idea or set of social relations (Weedon 1987). Ideology is thus what enables the capitalist system to function by masking the oppression of those upon whose work the economy depends. In this way, capitalism always operates in favor of the ruling class, who propagate dominant ideologies in order to maintain their own power and wealth in society.

While Marx sees ideology as primarily economic, French Marxist Louis Althusser (1969, 1971) defines ideology as that which masks reality within the social and political sphere. Althusser suggests that the individual is interpellated into subjecthood through institutional recognition, or the hailing process that takes place when we are categorized and named by structural forces, such as government, law, religion, family, school, and media. Through these apparatuses of the state, institutions impose on us a set of beliefs or way of being that limits what can and cannot be thought and done. "Social relation and processes are then appropriated by individuals only through the forms in which they are represented to those individuals" (Hebdige 1979, 13), generating a "false" view of reality that "represents the imaginary relationship of individuals to their real conditions of existence" (Althusser 1971, 153). As a result, Althusser (1971) suggests that ideology has no history, precisely because it appears as totalizing and incontrovertible, having sprung from nowhere and everywhere all at the same time.

These views underscore the ways in which girls' style is seen as a form of ideology. Girls have been repeatedly constructed as "clueless" in regards to consumerism. This understanding of girlhood comes from the commonsensical belief that girls are "natural" consumers "who derive an inordinate amount of pleasure from commodification and commodity fetishism" (Driscoll 2002, 110). This pleasure is not seen as a constructed truth that positions girls as supporters of the capitalist system, but rather as an inevitable part of who they "really" are. As a result, girls are depicted as the quintessential ideological victims who are easy targets for marketing ploys and the global capitalist machine. The main symbol of this cluelessness is fashion, an industry that has depended on the teenage girl consumer since its expansion into the mass market.

For Schrum (2004), the construction of girls as "natural" consumers begins much earlier than is generally thought: "In the consumer world, girls were the first teenagers. And they became the first teenagers in the decades before 1945" (4). While the teen market is usually thought to have emerged in the postwar economic boom, Schrum highlights its inception throughout the 1920s, 1930s, and early 1940s. During this time, girls emerged as a market that was distinct from boys. But Schrum suggests that gender differences did not drive consumer girls' culture so much as differentiation between age groups. While she acknowledges the impact that boys had on girls' identities, she highlights the demarcation of the teenage girl market from that of younger girls or college-aged women. As high school culture became more distinctive, teenage girls came to be viewed as a unique segment of the population, capable of making choices about how they wanted to dress and what they wanted to buy. As Schrum notes, the "growing awareness of high school girls as consumers with disposable income, influence over family purchases, and future status as adult shoppers" (19) pushed marketers and manufacturers to cultivate teenage girls as a niche market through age-specific advertising campaigns and commodities. This market was driven first and foremost by fashion, which was the earliest industry to recognize the potential of teenage girls' consumerism. "Local department stores were the first to experiment with teenage sections and with strategies for attracting and selling to high school girls" (5). By 1938, consumer girls' culture had emerged, bolstered by a market developed specifically around the words *teen* and *teenage*—"words that referred, without qualification, exclusively to females" (68).

In the postwar economic boom of the 1950s, consumer girls' culture helped to strengthen the capitalist system through the invention of the new teenage leisure market, which "transported adolescent consumption into a separate dimension of 'symbolic' and 'hedonistic' pleasures" (Carter 1984, 190). Consumer girls' culture became quite distinct from that of women, which focused almost exclusively on household "necessities" and the obligatory requirements of marriage and motherhood (Friedan 1963). Yet as distinct as these two markets were, girls and women crossed paths through increased public visibility in the new shopping centers and malls that developed around suburban lifestyles and the female-oriented marketplace. As Erica Carter (1984, 197) writes:

> Through their integration into an expanding teenage market, adolescent girls were drawn in increasing numbers into this new public space. The new generation of young consumers were particularly attracted to self-service and department stores, where they were free to look, compare and admire at their leisure, with no immediate compulsion to buy.

Malls became hangouts for girls, who could walk around freely, enjoying the company of girlfriends under the assumed supervision of adult store clerks and security guards. While girls learned some of the consumer skills they would need to be good wives and mothers at the mall, it was shopping that drew them out of the home and "into public life" (198). Girls used the mall as "a female substitute for the streets of male adolescence" (Lewis 1989, 100), where they were not necessarily positioned as sexual objects for teenage boys and older men. As McRobbie and Garber (1991, 5) note, "girls who spent too much time on the streets were assumed to be promiscuous." In the mall, a girl was free to hang out with her girlfriends, talk to boys without threat, and roam the myriad corridors of department stores in safety. Similarly, girls also spent a great deal of time in the home, where they prepared for their roles as wives and mothers (McRobbie 1991). In the home, girls were free to try on clothing with friends, to do each other's makeup, and to engage in forms of female bonding. Realizing this offshoot of consumer girls' culture, marketers and manufacturers continued to sell girls a "gendered cultural experience" (Lewis 1989, 89) in the home, where "there was room for a great deal of the new consumer culture within the confines of the girls' bedrooms" (McRobbie and Garber 1991, 6). The restrictions of these supervised locales linked girls even more tightly to consumer culture, filling such spaces with gendered products designed to enhance the time girls spent at home and in the bedroom.

As the capitalist system continued to organize "commodity markets to expand the boundaries of female consumption" (Carter 1984, 207), the growth of the fashion industry inevitably meant the growth of other related industries, such as diet, magazines, and makeup. The links between these industries further promoted consumerism as girls were bombarded with messages about how to look "better," dress "better," and be "better" people. If shorter skirts were "in," then magazines bemoaned the large ankle and marketed slimming cream; if sweaters were tighter, then the midriff had to be managed through new and improved undergarments; if girls wanted to attract boys, they were urged to buy "sexy" dresses. Marketers and manufacturers thus produced commodities geared toward "the female body itself" (205). But as Carter (1984) notes in her study of gender and consumerism in postwar Germany, these "improvements" to girls' physiques and styles were not for self-fulfillment or good health, but rather to increase girls' market value as attractive women. Girls did not just buy commodities, but were positioned *as* commodities in the competitive (meat) market of the dating world. She further emphasizes this point by drawing attention to Paul Willis' (1977) classic study on schoolboy "lads" and the sexism inherent in their working-class British culture. Carter (1984, 186) writes:

> In a story from which girls are largely absent, the moment of their appearance is profoundly significant. Girls, it seems, are written into youth cultural theory in the language of consumption; initially—though not . . . of necessity—as objects of consumption by men.

Through these discourses, consumption is naturalized as *the* defining feature of girlhood. Girls *buy* commodities *as* commodities,[3] a trope that has caused girls' culture to mean nothing more than a "culture of consumption" (Driscoll 1999, 177).

Positioning girls and women within a culture of consumption has been "crucial to recreating and maintaining a productive cycle that functions to support the capitalist economy" (Lewis 1989, 99). As Marxist feminists of the second wave understood, the gendered division of labor positioned women as the principal consumers, not just for the home, but also for those "little luxuries" that made the drudgery of their everyday lives palatable. This positioning was vital to capitalism and the industrialized economy. Marxist feminists worked to "wake" women up to this complicity, calling attention to women's consumer culture as nothing more than a tool of patriarchy that kept capitalism functioning (Barrett 1980; Mitchell 1984). A Marxist feminist critique of consumption thus asks, "whose needs are being served when femininity gives primacy to home and hearth and constructs women as consumers" (Currie 1999, 56)? Betty Friedan's (1963) "problem with no name" points to the false consciousness of women's household "busy work." Impossible to define or put one's finger on, the malaise caused by constant and empty consumption was not just how capitalism retained its primacy, but also how many women constructed their identities. Women came to understand themselves as "good" mothers and wives in relation to consumerism. As a form of ideology, however, this perceived false consciousness was critiqued by Marxist feminists as the source of women's subjugation and exemplary of the deluded lifestyle that they lived in suburban, middle-class America.

As essential as this critique was to an understanding of women's patriarchal oppression, it ironically served to diminish the everyday activities of women. In response, feminists working in the area of cultural criticism during the 1970s and 1980s sought to elevate rather than disparage women's consumer habits. Such reclamations focused on the idea of pleasure and fantasy in the reading of fashion magazines (Winship 1987) and romance novels (Radway 1984), and the watching of soap operas (Ang 1985) and films (Mulvey 1975). Shopping was also one of the distinctly female cultural practices that took on new meaning within this body of literature. Wilson (1985), for example, disputes the idea that fashion is nothing more than an ideological trap. She suggests that women's fashion cannot be explored without reference to power, fantasy, and self-expression. She also claims that the utilitarian approach ascribed to fashion by Marxist feminists suffered from the "inability to grant pleasure a proper place in human culture" (234). Further, she explains that if women are "fashion victims" (13), then so, too, are men, who must also buy items of apparel for work and leisure. Why should women be so ruthlessly critiqued for something men also regularly do?

Meaghan Morris (1993) writes lovingly of shopping, viewing the act of looking for clothes as a pleasant way to pass time and engage in the company of women. And Iris Marion Young (1990, 182) writes affectionately about the three pleasures she derives from the perusal of women's clothing: "touch, bonding, and fantasy," suggesting that it offers women the chance to know each other, unwind, engage in daydreams, and flirt with desire and sexuality—all the while exercising agency in their choices about how to present themselves in the social world. Such essays have gone a long way toward reclaiming women's consumption practices as important and interesting forms of self-expression in the public sphere, but where does that leave the consumption habits of girls, who are not included in these celebrations of shopping, clothing, and style?

As Gayle Wald (1998, 587) notes, the tendency in cultural theory has always been to "elide female subjectivity not only with consumption but with the commodity form itself." This trope has its roots in the Frankfurt School, particularly its flagship theory, the culture industry. Theodor Adorno and Max Horkheimer (1993, 120) suggest that the rise of mass culture produced a "culture industry," which "now impresses the same stamp on everything," confirming "the absolute power of capitalism." The culture industry is the very essence of crass commercialism, epitomized by easy and cheap production and a decided lack of creativity on the part of manufacturers and marketers. But the culture industry is also characterized by the "tasteless" consumers deemed to be ineluctably caught in a "circle of manipulation and retroactive need in which the unity of the [capitalist] system grows ever stronger" (121). Throughout the writings of the Frankfurt School, girls are exemplary of this manipulation, particularly in their tendency to incorporate commodities "into a false system of needs" (McRobbie 1997, 192). As Lisa Lewis (1989, 98) asserts, this view of girlhood, "has created an obstacle to the analytical consideration of consumer girl culture." As a result, cultural and feminist theorists largely ignored girls' cultural practices in the 1970s and 1980s, viewing them as a form of delusion rather than a complex strategical situation infused with power.[4]

The construction of girls' style as ideology also owes much to the rhetoric of marketing and advertising, where girls are heavily pursued as easily manipulated consumers. A successful marketing campaign is able to create a sense of identification and recognition between a product and its hoped-for customers. Through an Althusserian lens, this connection is created when marketers offer consumers "an entrance into the language of self ascription" (Butler 1997, 107) that enables them to identify with a particular commodity form. In this regard, a false consciousness takes over as we allow ourselves to become part of an imaginary world in which we feel intimately and personally positioned. From this perspective, girls are seen to be marketing victims par excellence, viewed as easily persuaded by the hail of advertising aimed at "suckering" them into stores to buy the latest tops, jeans, shoes, and accessories (McRobbie 1991). As always-already ideological subjects, girls, it is often thought, are readily convinced by advertising campaigns that make them think they "are the special 'you' of its discourse" (Storey 2006, 63).

In global capitalism,[5] the construction of girls as prototypical marketing victims has reached a feverish pitch. Consumer girls' culture entered the global era when, as McRobbie (2004, 5) suggests, "the concept of popular feminism found expression."[6] This expression manifested in the worldwide marketing of a product that both manufacturers and retailers felt would sell to girls in unprecedented quantities, otherwise known as girl power. Though girl power is a philosophy of empowerment first espoused by Riot Grrrl punk bands in the early 1990s, and, later, by the hugely successful pop group, the Spice Girls, in the late 1990s, marketers saw its potential as a commodity that could be sold in all forms, from lunch boxes to makeup to toys to cartoons, films, and television shows to style. In relation to style, girl power presents itself through the ideological construction of the empowered girl—a girl who is sassy, sexy, and strong. Emulating the sex appeal of the Spice Girls, girl power style includes T-shirts with slogans such as, "Girls Kick Ass," "Girls Rule, Boys Drool," and "Girls Rock," as well as decidedly sexual slogans where girls challenge boys and men to objectify them, such as "You Wish," "Dream On," "Slut," "Hottie," and "Make Me an Offer." Girl power style also includes other Spice Girls' favorites, such as platform shoes and boots, midriff-revealing tops, glittery makeup, push-up bras, miniskirts, and other glamorous accoutrement for bodily adornment. As extensions of the girl power trend, more ambitious endeavors ensued, including *La Senza for Girls* (pajamas, underwear, and "youthful" lingerie), *Cosmo Girl*, and *Teen Vogue*, all of which sell fashions marketed to an "empowered" girl audience.

As the "most prominent aspect of saleable Spice Girls authenticity" (Driscoll 1999, 178), girl power is marketed as a form of freedom, where girls can do whatever they want, be whomever they want, and buy whatever they want. As "a mass produced, globally distributed product" (178), girl power thus sells commodities laced with what is now widely regarded as a postfeminist ethos, so-called for its individualism, lack of collective political action, and focus on fashion, beauty, and body rituals (Gonick 2006; Griffin 2004; Harris 2004c; Taft 2004). For Ellen Riordan (2001, 291), girl power, as commodified feminism, does not empower girls at all, but, rather, stimulates "their dedicated consumption to pro-girl artifacts," especially those associated with fashion and style. Such artifacts, Riordan argues, "encourage girls' participation in passive consumption" (295).

As a marketing ploy, girl power has come to exemplify girls' consumer gullibility. One can almost hear the discussion in the boardroom: "Just slap a girl power sticker on that and they'll eat it up!" As Jessica Taft notes, girl power campaigns reinforce the idea that girls are easy marketing targets; all one has to do is invoke the proper ideological myth. Taft (2004, 74) explores the exploitation of this philosophy by quoting from an issue of *Fortune* magazine's featured article on girl power in 1997, at the height of Spice Girls popularity.

> Airshop knows about girl power. Its catalogue runs pictures and poems sent in by its creative customers. It publishes mini-profiles of teenage girls—girls who are kicking butt in male-dominated fields. . . . Is Airshop a hip, shoestring literary magazine or a catalogue for adolescent consumers? It's both. If you want to sell to the girl power crowd, you have to pretend that they're running things, that they're in charge.

Taft's analysis of girl power as a discourse shows that one of its key meanings is "consumer power," where girls are trained to think that the only power they have is "their purchasing power" (74).

On the international stage of economic exchange, girls' consumer habits are carefully cultivated as one of the pillars of global capitalism. This treatment can be seen in the explosion of "girl" oriented commodities in recent years, including television, film, music, videos, and fashions. Interestingly, style is the thread that joins most of these products together. Styles are sold through music videos and audience driven music shows, such as MTV's *Total Recall Live* and Much Music's *Much on Demand,* through reality television, such as *America's Next Top Model* and *Pussycat Dolls Present: The Search for the Next Doll,* where clothing makeovers are an integral feature of a glamorous transformation that takes place on the young female body, and through celebrities, who now have their own affordable clothing lines, such as Mary-Kate and Ashley Olsen and Gwen Stefani. Girls can also consume style through teen dramas that feature the latest, hottest trends worn by beautiful stars. The CW (formerly the WB), a television network that caters to teenage girls, features style as a main component of its Web site (http://cwtv.thecw/style). Girls can peruse styles worn by the cast members of any of its teen-focused shows under these browsable headings: brand, character, episode, product. As well, each show offers its own featured label of the week and a "what's hot" section for style. Intricately woven into most commodity forms, style is now the Rosetta stone of global capital's empire—easy to purchase, simple to market, inexpensive to produce, and effortless to display.

From early consumer girls' culture in the 1920s to the all-encompassing consumer girls' culture at the turn of the twenty-first century, clothing and fashion have remained essential to the formation of the teenage girl market and to capitalism. As a result, girls have been "repeatedly, and even obsessively, associated with the rise of mass culture" (Driscoll 2002, 11), where, as Kearney (1998, 287) notes, girls are often regarded as "passive victims of false consciousness caught up in the allegedly deceptive, alienating, and exploitive relations of commodity consumption." Given this construction, style is regularly seen as a symbol of girls' powerlessness in the face of an unrelenting capitalist system. But this powerlessness does not take into account the power that girls have as a sizeable and indispensable market in the global economy. As Carter (1984, 197) is quick to point out, girls "become visible at the point of consumption." And this visibility is power.

Teenage girls make up a vast market in North America. As Anita Harris (2004c, 166) observes, they are "imagined [by marketers] to have an enormous amount of control over family purchases, as well as considerable discretionary income of their own. For example, U.S. girls aged eight to eighteen are estimated to be worth $67 billion." The spending power of girls means that they are a much sought after demographic for marketers. In the competitive global marketplace, girls have more choices than ever before, ensuring that marketers and manufacturers must work harder than ever before to attract their attention. As Schrum (2004, 1–2) explains, by "the beginning of the twenty-first century . . . teenage fashions and trends supported a multi-billion dollar industry" where "teenagers spent $170 billion in 2002, about half of that on clothing." While these numbers include both boys and girls, Schrum points out that "girls are the primary spenders and the central advertising market for the majority of teenage products."

As consumer girls' culture is neither monolithic nor stable, marketers need to be aware of girls in multifaceted ways. While manufacturers, marketers, and storeowners depend on girls to purchase the fashions being retailed at the mall, girls also depend on marketers to offer them fashions that best reflect the characteristics they wish to emulate as they construct and negotiate their identities in the school. Marketers influence girls by positioning particular elements of style as "must have" trends. But girls also influence marketers by refusing trends, by demanding different kinds of attention and styles, and by acting in contradictory and inexplicable ways that make market research far from scientific. In Schrum's (2004, 20) analysis of early consumer girls' culture, she explains that girls did not simply buy anything and everything that was sold to them, but to some extent drove the market in new directions by expressing "strong interest in clothing and in developing teenage style." Driven in part by the growing teenage discourse, as well as an intensifying high school culture, where girls were free to form friendship groups outside of parental control, girls made demands on retailers to keep up with the times. But these consumer interests have always been somewhat elusive to marketers, who have never been able to predict just exactly what girls want. This lack of precision has propelled marketers to spend millions of dollars conducting research in order to gain insight into the "average" teenage girl, hosting focus groups and hiring "culture spies" to bring them this coveted information (Goodman 2003).

But consumer girls' culture is unpredictable and inconsistent. A popular fashion trend in one region or school may not catch on elsewhere; what flies off the racks at one store may hit the bargain bins at others. As a result, girls' style is not simply an ideological effect, where girls are unaware of how they are being marketed to or how they are being manipulated into buying particular items of apparel. Instead, style, as a point of convergence, causes girls and marketers to take notice of each other and to engage in relations of power that flow from marketers to girls and back again. Instead of being powerless to understand what is going on in consumer girls' culture, girls are vital to its definition and continued survival. Style is thus a threshold for contestation where girls exist in a steady stream of marketing influence, but they also exert influence as consumers who do not automatically and unconsciously buy what they are told. But as I point out in the following sections, the idea that girls are easily swept along by forces beyond their control plays a large part in how style means, not just as a form of ideology, but also as forms of conformity and pathology.

"APATHETIC": GIRLS' STYLE AS CONFORMITY

As a form of ideology, style is exemplary of girls' positioning within mass culture, where they are constructed as the "cultural dupes" who keep capitalism afloat. As conformity, girls' style is treated in a slightly different, though related, manner. Here, girls are denied access to style altogether, as style is viewed strictly as a form of subcultural resistance. Instead, girls are positioned within the realm of fashion, a term that is drained of politics and importance. While style is configured as a powerful form of engagement in the hegemonic arena of popular culture, fashion is configured as a powerless form of nonengagement in the conformist sphere of mass culture. This perceived lack of politics is signaled by girls' devotion to mainstream brands, stores, and labels that offer nothing more than trendy items of apparel for girls' pleasure-driven, indiscriminate spending. As I discuss shortly, unlike the "political" young men who use style to signify subcultural resistance, the association between girls and fashion leaves girls stranded in the culture industry, where their consumer choices are viewed as conformist, passive, and apathetic.

As conformists, girls are seen to be compliant with dominant norms and most comfortable when engaged in cultural practices that offer no risk or sense of differentiation. This construction is often typified by the female fan who not only fashions herself after her pop music idol, but also feels most satisfied when immersed in a crowd dressed exactly like her. From Beatlemania to Britney Spears, fandom is often defined as a hysterical form of passive submission, where screaming teenage girls gladly trade their individual identities for the security of looking like everyone else (see Ehrenreich et al. 1992). This common view of female submission is derived from several academic discourses that position girls within troubling binaries, including resistant/conformist, subculture/mainstream, producer/consumer, and political/apathetic. In each of these binaries, girls are cast in the disparaged terms (conformist, mainstream, consumer, apathetic), while boys and, to some extent, older teenage girls and young women are cast in the privileged terms (resistant, subculture, producer, political). In this section, I focus on how girls have repeatedly come to occupy the short end of these cultural sticks, particularly in relation to style.

Girls' cultural practices made their appearance—or rather nonappearance—in the work of British sociologists at the Centre for Contemporary Cultural Studies (CCCS) in the 1970s. If mass culture was decidedly apolitical, marked by the "pseudo-individuation" (Hodkinson 2002, 13) of mainstream female shoppers, then popular culture was *the* arena for creative cultural forms and resistance to the "official" culture of the dominant class (Cohen 1972; Hall and Jefferson 1976; Hebdige 1979; Willis 1977). While mass culture was viewed as passive, popular culture was active; while mass culture was compliant, popular culture was resistant; while mass culture was conformist, popular culture was creative; while mass culture was linked directly to the naturalized habits of girls who were enmeshed in consumer ideology, popular culture was associated with the "authentic" subcultural resistance of boys. Using Italian Marxist Antonio Gramsci's concept of hegemony as its theoretical base, the CCCS configured popular culture as the locale where items of mass culture were converted into "maps of meaning" for individuals or groups (Hall and Jefferson 1976). As Hall (1981, 239) famously notes, the realm of popular culture matters because it "is the arena of consent and resistance. It is partly where hegemony arises, and where it is secured." As a result of this theoretical framing, popular culture became a significant site of academic study, bringing with it a particular view of gender in relation to subcultural practices of resistance.

Not only was popular culture coded as male, but also, too, were the "heroic" figures featured in its exciting action narratives: punks, skinheads, rockers, Rastas, teddy boys, mods, bikers, and working-class "lads" (Cohen 1972; Hall and Jefferson 1976; Hebdige 1979, 1988; Willis 1977). The classic studies on subculture standardized the idea that "youth" meant "male" (McRobbie 1991). As feminist graduate students at the CCCS, McRobbie and Garber (1991, 4) thus ask if girls were really invisible in subcultures, or if they were merely eclipsed by male researcher bias. In typical accounts, girls and women are relegated to the sidelines, described as followers, hangers-on, and "slags" or "whores" (Willis 1977). This absence, McRobbie (1991, 25) suggests, unequivocally highlights "the redundancy of women in most subcultures."

Hebdige (1979, 133) defines subculture as "a form of resistance in which experienced contradictions and objections to ruling ideology are obliquely represented in style." For Hebdige (2), style is "a form of Refusal" that signifies a group's desire to separate from dominant society. It is through style that subcultures are first noticed and feared, as their experiences are organized and interpreted by the media. After Umberto Eco, Hebdige calls style "semiotic guerilla warfare" (105) in the sartorial struggle for differentiation and discernible opposition. In his classic book, *Subculture: The Meaning of Style,* Hebdige engages in "reading" the styles that Rastas, hipsters, beats, teddy boys, glam and glitter rockers, and, of course, punks took up. He asks, "how does a subculture make sense to its members? How is it made to signify disorder" (100)? In order to answer these questions, Hebdige analyses style through four categories: intentional communication, *bricolage,* homology, and signifying practices.

As intentional communication, style is viewed as a "visible construction, a loaded choice. It directs attention to itself; it gives itself to be read" (101). Using the example of punk style, Hebdige describes intentional communication as a display of subcultural codes, where styles are seen to be purposeful, not carelessly thrown together. Borrowing from anthropologist Claude Lévi-Strauss, Hebdige also analyses style as bricolage, or the unconventional use of conventional items. Engaged in "conspicuous consumption" (103), subcultures radically adopted commodities for purposes outside of their intended uses, making them appear both subversive and inventive. For punks, safety pins and dog collars became items that were separated from their original context and resignified through subcultural use. This collage look had shock value through the mismatched and ultimately jarring effect of everyday artifacts infused with profane, sick, or incomprehensible meanings.

Again borrowing from Lévi-Strauss, Hebdige also analyses style as homology, or the symbolic "fit" between style, music, values, and lifestyle. Styles chosen by particular subcultures were seen to be "either intrinsically or in their adopted forms, homologous with the focal concerns, activities, group structure and collective self-image of the subculture" (114). Punk style looked like vomit and spit, reflecting punk attitude and actions. Finally, Hebdige calls style a signifying practice. Rather than viewing style as fixed with meaning, he sees it as "in process," "active," "transitive," and "capable of infinite adoption" (119). In this regard, style is only ever rendered expressive through the signifying practices of a subculture's members, or how they use style to signify politics and to make meaning in their lives.

Given these complex configurations, for Hebdige and others at the CCCS, style was an enormously important component to subculture as it provided a crystallizing feature not just for its members, but also for those who researched and wrote about them. But Hebdige's mistake, McRobbie (1991, 24) suggests, is that he fails to highlight subculture's "best-kept secret, its claiming of style as a male but never unambiguously masculine prerogative. This is not to say that women are denied style, rather that the style of a subculture is primarily that of its men." Here, an interesting contradiction surfaces. In subculture, style is a meaningful cultural practice—but only for its male participants. As girls and women are ignored as subcultural participants, they are also denied access to the signifying practices associated with style. While male subcultural style "means," female subcultural style does not. Exploring this contradiction, McRobbie (1991, 25) notes that, "Hebdige's usage of 'style' structurally excludes women. This is ironic, for in mainstream popular culture it is accepted as primarily a female or feminine interest."

The invisibility of girls in the classic subcultural texts caused feminist youth cultural theorists to "search for autonomous female cultural forms" (Carter 1984, 187) in order to highlight the fact that they not only existed, but were also worthy of study. As a result, McRobbie (1991) decided to ethnographically explore working-class girls' experiences of culture.[7] She found "subtle" forms of resistance in the private spaces of the youth club, the home, and the school.[8] As opposed to the overtly antiauthoritarian practices of Willis' (1977) working-class "lads," McRobbie found that working-class girls "gently" resisted school culture, and that this resistance was linked to clothing, makeup, and fashion. Being "fashionable," McRobbie suggests, was in direct opposition to dominant school culture, as girls were indicating, "that they did not recognize the distinction between school and leisure" (44). Though McRobbie's language is hopeful and her data in depth, she admits that working-class girls' resistance was understated at best, focused specifically around a culture of consumption, where girls were still reinforcing dominant ideologies of femininity. Similarly, McRobbie and Garber

(1991, 8) analyze girls' participation in three subcultural forms: the motorbike girl, the mod girl, and the hippy girl. But in each of these instances, they also note that girls' participation, though visible and distinct from that of boys', nonetheless locates girls "nearer to the point of consumerism than to the 'ritual of resistance.'" They conclude that girls' visibility in these subcultures retains and reproduces their subordination.

In another oft-cited example, McRobbie and Garber (1991) focus on teenybopper culture as an example of a "distinctive" girls' culture that is "based round an endless flow of young male pop stars in magazines and pin ups" (11). Similarly, Simon Firth (1981) locates a unique female cultural practice in the form of "bedroom" culture. Bedroom culture exists in the private space of a girl's room, where she and her friends engage in heterosexual fantasies and beauty practices.[9] Yet, as Kearney (1998, 255) points out, both teenybopper and bedroom cultures "did not involve an investigation into other productive practices undertaken by female adolescents, and therefore ultimately reproduced the dominant notion that girls are capable only of cultural consumption." As well, for Holland and Eisenhart (1990), such exclusively "feminine" cultural forms show how girls are "educated in romance" in order to fulfill their "natural" roles as mothers and wives—roles punctuated by conformity to dominant femininity. Remaining optimistic but realistic, McRobbie and Garber (1991, 14) conclude that girls' participation in subcultural practices can only be understood "by moving away from the 'classic' subcultural terrain marked out as oppositional and creative by numerous sociologists."

Such studies positioned girls' culture within the mainstream, an ideological locale that is itself signified as "feminine" and "feminized" (Thornton 1997, 205). To be mainstream is to happily do what others do and to cultivate a lifestyle within the bounds of normative society, neither rocking the boat nor desiring difference. In relation to style, conformity means dressing like everyone else and gladly submitting to commonplace fashions that do not stand for anything in particular, other than one's dedication to following trends and remaining *en vogue*. Given this formulation, over the last decade and a half, feminist youth cultural theorists have endeavored to seek out "authentic" forms of girls' subcultural resistance, away from the "commodification and depoliticisation of girl culture" (Harris 2001, 133).

Though a number of examples have made their way into feminist cultural studies during the late 1990s and early 2000s (see Harris 2001; Kearney 1998), the most oft-cited example of girls' "authentic" subcultural resistance is the political network Riot Grrrl. Riot Grrrl signifies not just a form of music (all-girl punk) or a network of communications (countless zines and Internet sites dedicated to the dissemination of information), but also a form of "youthful feminism" (Garrison 2000, 143) committed to social change for girls and young women. Riot Grrrl band lyrics and zines were devoted to exposing the violence, sexual harassment, and sexism in girls' everyday lives, as well as the overtly patriarchal world of punk rock, especially live shows, where girls were relegated to the sidelines while boys and men moshed in the pit closest to the stage. The neologism "Grrrl" was significant in setting the stage for Riot Grrrl politics, as it invested the word "girl" "with a new set of connotations. It signified an angry, assertive feminist who relished engaging in activity" (Leonard 1997, 232). And just as Riot Grrrl bands were active in their staging and living of feminist politics, Riot Grrrl fans were also heavily ensconced in the network's pro-girl, anti-oppression attitude, starting up zines and websites dedicated to spreading "revolution, girl style."

While much has been written on the Riot Grrrl phenomenon,[10] what bears repeating for the purposes of this discussion is the use of "female display" (Leonard 1997, 235) in Riot Grrrl politics. Riot Grrrl bands wrote "slut," "rape," "shame," and other words typical of female oppression on their bodies in black marker. Band members also sometimes dressed in specifically feminine ways, wearing baby doll dresses with "flirty" floral prints, and intricate hairdos. For Marion Leonard, the "critique of the construction of the feminine is abundantly clear when this attire is juxtaposed with bold words such as 'whore' written on the body" (236). Riot Grrrl fans emulated this "cut-up" style, using juxtaposition to draw attention to feminine norms and bodily restrictions. Retro 1950s dresses were juxtaposed with combat boots, punk haircuts were juxtaposed with hot red lipstick and eye makeup, and camouflage was juxtaposed with platform and high heel shoes (Klein 1997). These styles challenged the very ideas of girlhood, femininity, and heterosexuality while remaining "true" to the classic punk ethos of DIY (do-it-yourself). Riot Grrrl bands and fans thus used style as a key mode of resistance that spoke volumes about girls' positioning in North American society.

Repeatedly emphasized as one of the few "authentic" subcultural forms in which girls have been engaged, Riot Grrrl was constructed as a nonconsumer movement that successfully remained on the fringes of mainstream culture by instituting press blackouts and refusing to allow mass media to co-opt their feminist message, music, and style. In this respect, Riot Grrrl "authenticity" is often juxtaposed to Spice Girl "inauthenticity." Constructed as the death of feminist politics (Bellafante 1998; Labi 1998), the Spice Girls are seen to be the ultimate commercialization and commodification of Riot Grrrl resistance. As Gonick

(2006, 10) notes, girl power's popularity "is credited to its very lack of threat to the status quo for the ways in which it reflects the ideologies of white, middle-class individualism and personal responsibility over collective responses to social problems." Seen to offer no "real" politics to speak of, the Spice Girls were also constructed as exemplary of postfeminist apathy and girls' indifferent view toward social change.[11]

The "authenticity" of Riot Grrrl and the "inauthenticity" of the Spice Girls points to another significant dichotomy in relation to how girls' style means as a form of conformity. Not only are girls held in binary tension with subcultural boys, but girls are also held in binary tension with young women who are described as mainly white, middle-class, and university-educated. Such girls have greater access to the Riot Grrrl network through their ability to go to live shows and log on in the privacy of their own rooms. While the Spice Girl fan-base is known to be younger teen and tween girls[12] who need parental permission to attend concerts, the Riot Grrrl network included mainly older teenage girls and young women, with ages ranging from fourteen to thirty, augmenting the term "youth" in descriptions of its participants (Leonard 1997). As Driscoll (1999, 181) convincingly suggests, though Riot Grrrl bands could have provided many girls with access to a political stance and an oppositional language, far "fewer girls are likely to have a space to listen to Bikini Kill than to the Spice Girls."

The construction of "authentic" and "inauthentic" girls' culture suggests that only certain girls have entrée into the hegemonic world of popular culture, where girls can engage in practices of power through fandom, cultural production, gendered communities, and style. But such a distinction is predicated on a false dichotomy not just between "authentic" and "inauthentic" cultural forms, but also between resistance and conformity generally. As Kearney (1997, 209) suggests, radical youth cultures, such as Riot Grrrl, do not "exist in some 'free' or 'authentic' state before the mainstream media become aware of their existence." Subcultures depend on elements of mainstream style in order to showcase resistance through the resignifying practice of bricolage. The media then pick up on these styles, converting them back into mainstream culture through co-option and appropriation in the marketplace. Subcultures then continue to resignify these same elements of style, viewing them, once again, as symbolic of conformity and mass production. And the cycle continues. Sarah Thornton (1994, 176) makes this point in her analysis of acid house and rave music. Subcultures, she writes, "do not germinate from a seed and grow by force of their own energy into mysterious movements to be belatedly digested by the media. Rather, media are there and effective right from the 'start.'" In other words, there is no purely "authentic" cultural form that has not been infused with or influenced by mainstream culture.

I am not suggesting that Riot Grrrl and the Spice Girls exist on the same cultural trajectory. Of course, there are differences between these two versions of girl power and the range of styles that fans of both use to show affiliation and belonging. But what I am suggesting is that "authenticity" and "inauthenticity" are not diametrically opposed; rather, they depend upon each other for mutual definition. As a result, each term supports and draws strength from the other rather than existing in purely differentiated spheres. Mainstream girls' culture—viewed as conformist—is the very basis upon which girls' subcultures—viewed as resistant—revolve. Neither they nor their styles can exist in a vacuum, but they draw continuously and fluidly from one another, diminishing the distance between them.

Girls' style is viewed as conformist because girls have been constructed as followers and copycats in the classic subcultural texts, where girls were assumed to "simply follow the same cultural trajectory as boys, but with less involvement, commitment or investment" (McRobbie 1991, 35). Girls' style means as a form of conformity because attempts to highlight its visibility in uniquely female cultural expressions have inevitably shown girls' participation to revolve around seemingly conventional modes of consumption. Girls' style also means as a form of conformity because of a new dichotomy that has been constructed between younger teenage girls who have limited access to alternative cultures, and older, more educated young women, who have greater chances for encountering and embracing forms of feminist politics. While the latter group is often viewed as a powerful producer of culture, the former is often disparaged for "merely" consuming.

Given these complex articulations, it is worth considering whether or not conformity through style is really possible, as each girl, each body, each identity is always "in process" and on the move. While mainstream styles are, by definition, the most popular and, therefore, the most sought after forms of dress in our society, girls can never wear styles in exactly the same manner. As each girl constructs and negotiates her identity differently, so, too, will each girl wear her pants, shoes, and tops differently as a form of social skin. While a certain amount of conformity (or uniform-ity) is crucial if one is to become part of any group—subcultural or otherwise—within modes of dress that enable members to recognize each other as the same, overt and covert differences inevitably exist. One has only to think of the schoolgirl uniform as an example. On first glance, these kilted students all look the same. But upon closer examination, it is easy to spot the differences among them. One girl rolls her skirt up half an inch higher than the others. One girl wears darker eye-liner than the others. One girl wears platform boots. One girl wears pink shoelaces. One girl sports a faint streak of purple in her hair. One girl has a double-pierced earlobe. One girl wears a gold cross, dangling on a chain. One girl's blouse is perfectly pressed. And so on.

While girls' mainstream style is often looked down upon for being conformist, as I noted in chapter 1, any form of style—resistant or not—is an engagement with identity. This engagement is not only a function of how structural forces, such as gender, race, ethnicity, class, and sexuality, position girls but also a function of how girls actively work to reposition themselves within these seemingly stable categories. As I show in chapters 4 and, particularly, 5, girls engage in relations of power when they use style to actively negotiate how others see them. This nego-tiation need not be resistant or even overt in order to be influential in the lives of everyday girls, making "mainstream" style as powerful as "resist-ant" style in the constant struggle for social intelligibility. Conceived in this way, style is powerful as it is one of the most obvious and accessible ways that a girl can gain some measure of control over how others see her and how she sees herself. Turning next to the third dominant meaning of style, this power becomes even more significant when viewed through the lens of developmental psychology, where girls are not only denied access to resistant styles, but to untraditional girlhoods as well.

"DESPERATE": GIRLS' STYLE AS PATHOLOGY

I first encountered Lauren Greenfield's (2002) photographic essay on girls when I read a review in the local newspaper. The towering headline read: "Lauren Greenfield's portraits reveal the disturbing hunger, the need to be desired, even objectified, that typifies the lives of ordinary girls" (Woodend 2002, H8). When I later came across the large coffee-table book in a bookstore, I marveled at its evocative cover. It depicted two fifteen-year-old girls in a department store change room. Sheena—in heavy black eyeliner, a dangling silver cross necklace, and florescent blue nail polish—is scrutinizing her cleavage by squeezing her breasts together in a tiny purple shirt that looks more like a bra than a top. Her expression betrays her as we see the dissatisfaction in her eyes. She is unhappy, agitated. Amber, her friend, sits on the floor of the change room, presum-ably offering fashion advice and the kind of support that girls so often give to each other when shopping. But no amount of support will placate Sheena and make her believe she looks good in that top. As she stares into the change room mirror, we see the frustration and contempt she holds for herself and the desperation with which she squeezes her breasts together in order to make her cleavage look ample enough—perhaps to impress the boys at school, perhaps to impress the girls, or both.

Flipping through the book, it is easy to see how Greenfield defines *Girl Culture,* the title and theme of her collection. Girls' culture[13] is tanning booths, beauty pageants, fat camps, fashion shows, beauty rituals in bathrooms, pop star adulation, ritualistic exercise, topless competitions during Spring Break, bikini contests, strip clubs, cosmetic surgery, and shopping. Most, if not all, of the images are hauntingly sad—infused with glossy emptiness and posed girls who look off into the distance, muted and constrained. Girls are depicted as struggling with anything and everything, including self-mutilation, anorexia, narcissism, weight loss, sexual harassment, objectification by men, the overwhelming desire to be beautiful, and their obsessive need to be looked at and adored. Greenfield's images depict girls' culture as a prison of peer and media pressure, with no hope for escape. Peer pressure is shown to encompass social demands from girls, boys, and parents, as the textual blurbs that accompany many of the images highlight. And media pressure is shown to encompass just about everything else, from pornography to music videos to fashion magazines to Las Vegas showgirls to the Playboy Mansion. Media pressure also emanates from the adult images of femininity and sexuality to which girls are regularly exposed, causing Greenfield to include photographs of celebrities in the collection, such as Cindy Margolis, Jessica Alba, Jennifer Lopez, Gwen Stefani, Venus Williams, Taylor Wayne (a pornographic film star), and Aki (the star showgirl at the Stardust Hotel, Las Vegas).

Joan Jacob Brumberg's (2002) introduction to the collection offers further explanation as to what Greenfield means by girls' culture. She highlights girls' cultural practices, as they existed in North America during the 1900s, suggesting its roots were in "family, school, and community." Brumberg notes that girls of that era,

> chattered about new hair ribbons and dress styles and inscribed sentimental rhymes in one another's autograph books. In private, many prayed and wrote earnestly in their diaries about how they wanted to improve themselves by helping others or becoming more serious people. (Para 2)

In sharp contrast, today's girls' culture, Brumberg writes, is "driven largely by commercial forces outside the family and local community. Peers seem to supplant parents as a source of authority; anxiety has replaced innocence" (para 3). As she suggests later on, today's girls' culture is infused with a "tentative, diffident, even desperate" (para 12) flavor of insecurity and low self-esteem.

Girls' style is seen as a form of pathology precisely because of such images, where girls are depicted as sad, frantic, and distracted by their looks—overwhelmed by the pressures of shopping and bodily adornment. This view of style is part and parcel of a broader discourse on girls that I call the Ophelia genre, an academic and professional body of literature that frames girls' cultural practices as symptoms of low self-esteem.[14] Named for Mary Pipher's (1994) wildly successful book, *Reviving Ophelia,* the Ophelia metaphor is based on the Shakespearean heroine who drowns herself over the loss of Hamlet's affection and her father's constant dissatisfaction. Driven to self-hatred by this lack of approval and her inability to please any of the men in her life, Ophelia goes mad and commits suicide. Like the tragic character in *Hamlet* and the tragic characters in Greenfield's photographic essay, Pipher views teenage girls as helplessly tossed about by external forces that dominate their lives, such as "junk" culture, peer pressure, and developmental changes that turn "normal" and "happy" preadolescent girls into "miserable" teenagers who engage in desperate behavior in order to fill the emotional void they have suddenly developed. This void, Pipher suggests, is the result of being forced to conform to a "fake" self that serves to please others. As a result, girls begin to feel depressed and act "overwhelmed and symptomatic" (13) in their "false" identities.

Telling stories from her own psychotherapy practice, Pipher questions our media-saturated culture, blaming it for the bulk of girls' troubles.

> Why are so many girls in therapy in the 1990s? Why are there more self-mutilators? *What is the meaning of lip, nose and eyebrow piercings?* How do I help thirteen-year-olds deal with herpes or genital warts? Why are drugs and alcohol so common in the stories of seventh-graders? Why do so many girls hate their parents? (11, emphasis added)

Intriguingly, Pipher includes elements of style in this list. Are lip, nose, and eyebrow piercings signs of self-mutilation, herpes, and drug use? In the case studies that fill the book, how girls dress is linked to their unstable behavior, such as the case of Holly, who "was a Prince fan and wore only purple. Her father brought her in after a suicide attempt" (21); and Gail, who "burned and cut herself when she was unhappy. Dressed in black, thin as a straw, she sat silently before me, her hair a mess, her ears, lips and nose all pierced with rings (21); and Charlotte, who "wore seductive clothes, dyed her hair Madonna blond and did whatever she pleased" (47).

In the Ophelia genre, girls are represented as desperate for love and connection, not only unable to navigate our "damaging" culture, but also unable to ask for help. As Pipher suggests, one, "way to think about all the pain and pathology of adolescence is to say that the culture is just too hard for most girls to understand and master" (13). This lack of understanding and mastery is exemplified, according to Pipher, by elements of style, such as piercings, tattoos, hair dye, subcultural paraphernalia, and "sexy" fashions. Such items are seen to be illustrative of girls' sinking self-esteem and symptomatic of insecurity, unhappiness, and depression. According to Pipher, they also showcase girls' inability to resist "junk culture" and peer pressure; desperate to fit in, girls will do and wear just about anything.

Brumberg (1997), author of *The Body Project,* suggests that girls' cultural practices in the twentieth century are indicative of a historical shift away from "good works" toward "good looks." She explains that things are getting much worse for girls as our media-saturated, sex-soaked culture compels them to spend more and more time patrolling the borders of their bodies in order to feel comfortable with themselves. From dieting to bikini waxing to menstruation, from shopping to piercing and tattooing, the body, Brumberg writes, has been made into "an all-consuming project" (xvii) for girls today:

> A century ago, American women were lacing themselves into corsets and teaching their adolescent daughters to do the same; today teens shop for thong bikinis on their own, and their middle-class mothers are likely to be uninvolved until the credit card bill arrives in the mail. (xvii–xviii)

For Brumberg, the highly "individualistic" activities of today's girls' culture pales in comparison to the more communal cultural practices of the past. Shopping and trying on clothes are viewed negatively, indicative of girls' narcissistic lifestyle and their declining social values.

For Greenfield, Pipher, and Blumberg, the nature of girls' cultural practices implies that girls today have it much worse than girls of any preceding era. They base this opinion on the presence of media in girls' lives. Everywhere girls look, there are images of female icons with perfect bodies in perfect clothes. Pipher, for example, compares her own "innocent" girlhood in the 1960s to girlhood today, as it develops within our gloomy media culture. Pipher (1994, 12) asks: "Why are girls having more trouble now than my friends and I had when we were adolescents? Many of us hated our adolescent years, yet for the most part we weren't suicidal and we didn't develop eating disorders, cut ourselves or run away from home." As media pressure has increased, she asserts, so, too, have girls' problems.

This perspective is starkly represented in the 2003 film, *Thirteen,*[15] where Tracy Freeland (played by Evan Rachel Wood) undergoes a radical transformation over the span of three months that takes her from "good" girl to "bad" girl right under her mother's nose. Cowritten and directed by Catherine Hardwicke, *Thirteen* chronicles Tracy's desire to transcend her current social status as a nice, smart, but unpopular girl in order to become friends with the beautiful, popular, and unstable Evie (played by coauthor Nikki Reed). In order to prove herself worthy of Evie's attention, Tracy must steal, lie, drink, huff aerosols, take hallucinogenic drugs, and engage in sexually mature situations. As perilous as these activities may be to Tracy's mental and physical well-being, she manages to accomplish them in order to become Evie's best friend.

Becoming Evie's best friend also necessitates a shift in style. In the film's opening scenes, Tracy wears "regular" jeans and T-shirts, sneakers, and no makeup or accessories. Instead of a glamorous and "sexy" teenage girl, Tracy looks wholesome and plain. But after Evie and Tracy become inseparable, her style starts to change. Tracy begins wearing midriff-revealing tank tops, low-rise jeans, thong underwear that peaks above her waistline, miniskirts, tube tops, halters, and heavy eye makeup. Under Evie's influence, Tracy also pierces her navel and tongue, becoming someone that even her own brother, Mason, does not recognize. In low-rise jeans, a bare midriff, a tiny little top that resembles a bra, and teased, wild hair, Tracy is shown suggestively leaning over a counter in a video arcade on Melrose Avenue, when Mason and a group of his friends walk by. Mason's reaction to this "girl" causes him to call out, "Oh, baby, back that ass up!" before realizing it is his sister. And in one of the film's most emotional scenes, Mel finally realizes that her daughter is "in trouble" when Tracy taunts her with this description of her new style: "No bra, no panties! No bra, no panties! No bra, no panties!"

With the tagline, "It's happening so fast" and the alternate, "They're not little girls anymore," *Thirteen* is a morality tale about the consequences of girlhood out of control. Tracy cuts herself in the bathroom, drinks herself sick, drops acid in the park, and gives blowjobs to acquaintances. Set to a backdrop of parental neglect, peer pressure, and media images that bombard girls with messages about how to be "sexy," *Thirteen* suggests that girls today are in deep trouble. Forced to give up their former, preadolescent selves, girls must take up "new" identities that grant them acceptance in the school. While this transformation is shown through a changing personality, erratic moods, and risky behavior, its most visible characteristic is a change in style.

Thirteen is a terrifying film for parents, as it suggests that media and peers overwhelmingly influence girls today, and that this influence makes girls feel empty enough inside to hurt themselves or worse. Pursuing popularity and sexual attention, the film suggests that girls willingly let go of any meaningful relationships that might offer them emotional sustenance during the "troubled" teen years. The film also condemns parents, particularly working-class single mothers, for not giving their daughters enough attention. In the film's dramatic conclusion, Mel learns just how out of control Tracy has become, oblivious to the depths that she had sunk.

> *Mel:* How do you explain $860 in your wallet?
> *Tracy:* We jacked it, okay? God, Mom, you knew what was going on with all those clothes and shit. Not even you're that dumb—
> *Mel:* [shouts] I didn't know it went that far!

In the end, the film's unpleasant moral is that girls lack the necessary tools to safely navigate the treacherous cultural changes that have made our society a "girl destroying" (Pipher 1994, 12) place over the last two decades.[16]

In *Thirteen,* style is a form of pathology, as well as ideology and conformity. As pathology, style is indicative of Tracy's mental instability and acts as a visual representation of her depression, cutting, sexual "deviance," and out of control behavior. While Mel is depicted as unaware of her daughter's downward spiral, Tracy's style continuously acts as a cry for help that should have been impossible to ignore. Her shift in clothing is illustrative of her desperation to: fit in, become popular, gain the favor of boys, fill the emptiness inside, and win back her mother's attention. As a result, the film also depicts style as a form of ideology, wherein Tracy is easily swayed by media messages that tell her how to dress like a "sex pot," and what she must buy in order to do so. Finally, the film shows style as a form of conformity, as Tracy willingly chooses to lose herself in the popular crowd by dressing in a way that causes her to become just like her newfound friends.

Style as pathology relies on the idea that girls develop in a linear and natural manner that moves from girlhood to youth to adulthood in an inevitable progression. Inherent in this account is the idea that identity is unwavering and coherent. Any move away from this developmental path is viewed with psychological concern and as a departure from one's "true" self. But as Gonick (2006, 13) is quick to point out, representing female adolescence as chaos "feeds into many of the demeaning cultural stereotypes about girls and young women. Girls' behaviors and emotions are represented as though they are beyond the bounds of comprehension." Helpless to help themselves, girls are represented as victims, without a clue and without the wherewithal to stand against the crowd. As ideology, conformity, and pathology, style thus signifies girls' inability to live emotionally healthy lives and, instead, offers "proof" that girls require more scrutiny, concern, and discipline than boys. As Harris (2004a, xxiv) suggests, this growing desire to "analyze young women has often concealed another agenda: the regulation and surveillance of their behavior."

While particular styles are often symbolic of psychological distress or warning signs for emotional trouble, this powerless construction ignores all that is pleasurable, productive, and useful about girls' style, particularly in the lives of girls who, for one reason or another, feel like outsiders. For many girls, style is a form of fun and fantasy, freedom and autonomy. Rather than a cry for help, style can be a lifesaver, enabling girls to engage in a form of self-expression that offers them some sense of control. Girls also use style as a feature of socializing, enabling them to discuss how they want to look and dress, while trying on clothes and spending time together. During these moments, spaces can open up for related discussions of body image, constructions of gender and sexuality, advertising and marketing, the production of trends, and the pressure girls feel to look and dress a certain way. Such conversations also have the power to create rapport, friendship, and community, as well as the potential to develop critical thinking in relation to style. While not all girls enjoy talking about or buying clothing, it can offer those who do the chance to forge intimacies, strengthen relationships, and engage in conversations about who they are and how they want to be perceived in the social world.

Style as pathology also negates the power of fantasy in girls' lives. Through style, girls can talk about their desire to dress differently and test out new ways of engaging in embodied subjectivities. In such moments, girls may recognize different ways of being or gain approval for an idea that they have been toying with in secrecy. Girls can also use style to flirt with desires, sexualities, and lifestyles without having to give them public voice. In the imagination, girls can try on outfits that they would never wear to school for fear of ridicule. Style opens the door for fantasies that might otherwise seem impossible, dangerous, or intimidating. Alone in her room, a girl can also use style similarly, enacting performances of girlhood in clothing borrowed from friends, siblings, and parents. It is here, in private space, where girls might gather the courage to debut a new image at school, practicing how to look and be in radically different subject positions.

The Ophelia genre fails to recognize these powerful uses of style in girls' lives and instead focuses on a narrow definition of girlhood that suggests girls must act and dress a certain way in order to be "healthy." Piercings, hair dye, masculine and subcultural looks, and overtly "sexual" styles position girls away from traditional femininity, where "girls are girls." Instead, the boundaries of girlhood become blurred, complicated. While this blurring may prompt some developmental psychologists to see girlhood as "off the rails," for others, this move away from a femininity marked by white, middle-class, heteronormative values is a powerful way to decenter and resignify girlhood itself, pushing it in new and exciting directions, where the word *girl* comes to be inscribed with more and more meanings through more and more styles. While I agree with Pipher that we certainly live in a culture that does not concern itself with girls' well-being, style may, in fact, offer girls a way to deal with some of the harsh realities that they face on a daily basis. For some girls, style is a savior, a best friend, and a shelter from the misery of peer and media pressure. For some girls, style is a secret pleasure that enables them to get through the day. For some girls, style is libratory.

Girls are indeed targets of marketing, global capitalism, media, and professional prescriptions. As a point of convergence between girls and power, however, style acts as a form of contestation, where girls are both influenced by these forces and influencing at the same time. While girls are influenced to buy certain items of apparel through capitalist ideologies, they also drive trends and shape consumer girls' culture. While girls are denied style with semiotic significance and, instead, are stranded within the apolitical world of fashion, mainstream styles are inextricably linked to subculture, the market, and the play of bricolage. And while girls are bombarded with damaging and ruthless images that make them feel badly about themselves, girls are also resilient, smart, and savvy. While some girls may need help navigating the injurious terrain of media, others may ignore it, scoff at it, or (like most of us) live in ambivalent tension to it, sometimes feeling badly, and other times rising triumphantly above the fray. Girls may use style to gain power over oppressive forces, hugging it close and nurturing it like a trusted confidant that enables them to keep the howling winds at bay (to use Pipher's metaphor).

Style offers girls access to these multifaceted processes as a complex strategical situation that enables them to engage in relations of power. But as I noted at the beginning of this chapter, a girl's engagement in relations of power does not negate the influences of sociocultural forces in her life—it merely complicates them, suggesting that there is more to girls' style than dominant meanings of ideology, conformity, and pathology imply. But as I turn next to the school and girls' stories of identity negotiations through style, it is useful to keep these dominant meanings in mind. They help to contextualize girls' ambivalent desires and constant struggles for control over how they were seen by others. Girls at ESH regularly spoke of "evil" marketers," media images that were indelibly etched in their minds, like a "tattoo that you can't get off," and the general impression that adults seemed to have of them as girls. Such influences do not fade from the scene as we enter the school; they become absorbed into the air that girls breathe and the identities that girls live.

CHAPTER 3

"WHERE IT'S AT": RHETORICAL PERFORMANCES OF EAST SIDE HIGH

It's pretty much an "in school" thing. You know what I mean?

Isabel, "dressy" style

WHEN I ENTERED MS. RIPPLE'S GRADE EIGHT SCIENCE CLASS, it was abuzz with after-lunch excitement. Girls barely noticed my presence as I sat at an empty stool behind a dull black lab table covered in tag graffiti and liquid paper art. After Ms. Ripple gave out the instructions for the day's lab experiment, I turned to the two girls on my left and introduced myself. Ling and Ana were only too happy to have a diversion from the work at hand. They were self-described "quiet Chinese girls" who wore loose fitting jeans, T-shirts, hoodies, and nonlabel sneakers. They did not wear any makeup and both had their hair pulled back in neat ponytails. After a brief explanation as to what I was doing in their school, I took the opportunity to ask them the question with which I had been preoccupied since arriving at ESH six weeks ago: "Can you describe the social groups here?"

Ana started to talk excitedly. "We discuss this topic *all* the time," she replied. "Yeah," Ling said, cutting her off, "we discuss it all the time. Um, it's important to know that it's *not* like what you think here." When I asked what she meant, Ling explained, "It's not like the movies. There's not one major popular group." Ana nodded. "We thought it would be like something else, but it's not." I asked what they thought it would be like. Ling responded first: "Um, like Hollywood movies where there's like, this group of bitchy girls who make everyone's life miserable."[1] Ana added: "Everyone always talks about that like, at other schools, too. There's always this group of girls who 'rule the school,' but at ESH, it isn't like that." "No," Ling said. "It's nicer here." After a minute or two of paying attention to Ms. Ripple, Ling leaned over to me again: "All groups are even," she whispered. "You can find your own people."

While I did not say so, I was surprised to hear Ling and Ana's utopian version of ESH. I had already met lots of girls who felt ostracized for various reasons. Diane had told me that the girls in French Immersion were ruthlessly cliquey and exclusionary. I also knew that hierarchies existed at the school. Each curricular track had its place in the pecking order, ranging from advanced academic programs at the top to the Aboriginal program at the bottom. I also knew from my wanderings around the school that there were lots of students who sat alone by their lockers at lunch, not yet having found "their own people." But Ling and Ana's view also made sense. The highly decentralized and fragmented social world created by multiple racial, ethnic, and language groups, curricular tracks, and lifestyle affiliations made it seem possible to believe that a group existed for everyone out there in the wild profusion of the school. But this view of ESH was certainly not the only view. I had gathered numerous descriptions of the school's social scenes, groups, and cliques[2] in the short time that I had been at ESH, all of which seemed to contain gaps and inconsistencies.

Dren told me that the social world at ESH included the "Asian gang groups" and "all the Asian girls who think that social life in high school is like, the most important thing in the world." It also included the students who "hang out and smoke like, down by the car area." She completed her description by adding in "the 'bad-ass' kids" who start "girl cat fights." When I asked her about other social groups, she took a moment to think and then responded: "And then, you have, like—I don't really pay attention to the rest of the groups that much."

Adrianna described ESH's social world this way: "There's um, there's the Frenchies—French Immersions, Accelerated Studies, Independent Learners. And we have all the Kappa[3] people." Adrianna took a moment to think. "Um, there's also a Spanish group of older people, and then uh, there's trading card collectors, you know Yugio cards? That's in the library." Shen began her description of ESH by first listing the skaters, defining them as "Caucasian," then "those sort of gang member type people, who think they're cool and dress like idiots!" She separated these "extreme people" from "the people who are like, sort of normal, who are a bit of everything." She then lumped together the various advanced academic programs: "Yeah, there's the smart people who are sort of nerdy

who have no fashion sense who don't fit in and who are ignored by people and bothered and picked on." Finally, she concluded by listing some
of the racially organized groups: "There's the hardcore Chinese guys,
like those little Kappa people. And those, like, Spanish-type people."

Marla mapped out her version of the school on a piece of loose-leaf
paper that she ripped out of her three-ring binder. Her map divided
the school into buildings, portables, the parking lot, smoking areas, and
locker banks. She placed the skaters at one end of the school, writing
"hot guys" beside their name, and the Nammers or Vietnamese group in
another corner, writing "cool kids" beside their name. Then she added
her own social group on the third floor, writing, "grade ten Frenchies."
I asked her about the rest of the school, surprised at how few groups she
had listed. "Oh, I forgot," she replied. She added in the "drama kids" in
the drama room, the Hispanic group in the "Spanish Hall" and then
listed the "special needs kids" in one wing of the school, writing "bad
asses" beside their name. Finally, she included the Aboriginal program in
the basement under the title "gangs." When she was done, she proudly
handed me the map and said, "there you go." "That's it?" I asked. "Yeah,
that's it," she replied.

As a school ethnographer, these partial and contradictory descriptions
initially frustrated me. I poured over my field notes and bits of interview
data, attempting to "piece" the school together. I made lists at the back
of my notebook in order to keep everything straight: Nammers (Vietnamese), Hongers (Hong Kong), Flips (Filipino), Frenchies (French Immersion), skaters (carried and/or used skateboards), drama kids (hung
out in the drama room at lunch), Accelerated Studies (advanced academic program), Independent Learners (self-directed academic program), beauty school girls (Aesthetician program), Hispanic kids, kids in
various special needs programs, smokers (hung out in smoking areas/
cross-listed with beauty girls), art kids (hung out in the art rooms at
lunch), Mr. Weston devotees (hung out in Mr. Weston's geography classroom at lunch), computer kids (hung out in the computer lab at lunch),
trading card collectors (spent lunch trading Japanese anime cards outside of the library), CBCs (Canadian Born Chinese), ESLs (English as
Second Language), and numerous other social groups that came in and
out of view depending on who I was talking to and where I happened to
be sitting at the time.

My initial desire to "map" the school stemmed from the fact that mapping schools is what school ethnographers do. It is our job to make the
school intelligible to readers, to persuade them of its "realness," and to create
a sensory experience that is almost as good as being there. The reader's
ability to buy what the ethnography is selling is crucial to its success. Such

a rhetorical struggle begins first and foremost with the ethnographer's description of the school. The school is the stage upon which all the action takes place. It is the institutional glue that binds the identities of the students together. But it also has an identity of its own that shapes and colors everything that occurs within it. And just as the identities of girls are contextual, so too is the identity of the school. As Britzman (2000, 32) notes, every telling of the school "is constrained, partial, and determined by the discourses and histories that prefigure, even as they might promise, representation." Mapping the school is thus a forced event, offering what Foucault (1972, 35) calls a "false unity" among dissimilar and contradictory perspectives. In the smooth telling of ethnography, schools are organized around descriptive statements that promise cohesion and harmony. Myriad stories become one story—a story told through the eyes of the ethnographer, the cartographer who charts the way.

Such renderings make the school believable to readers, who can then imagine themselves wandering the halls as the story unfolds. But in Britzman's (1991) ethnography of student teachers, she comes to realize that these easily rendered versions do not do justice to how schools produce complex and multiple subjects who, in turn, produce complex and multiple schools. Britzman (2000, 30) offers "a more complicated version of how life is lived" by tracing, "but not without argument, the circulation of competing regimes of truth" for her research participants. Similarly, girls' inconsistent stories of ESH represented the school as it existed *for them,* but those understandings were based on girls' cultural locations, their positionings within the school's social world, and their own perceptions of what their school was all about. Girls in French Immersion had a radically different understanding of the school than girls in Accelerated Studies. Beauty girls had a radically different understanding than did girls in the Aboriginal program. Goth girls saw things in radically different ways than did preppy girls or punk girls or sporty girls or smokers. But within these social categories, there were still countless perspectives. Hearing similar stories of the school only meant that the girls doing the telling occupied parallel social and cultural locations and, even then, differences in perception were evident. To "map" the school would have forced me to either privilege some girls' views over others or to collapse and condense contradictory accounts into a false unity. But following Britzman (2000, 32), I felt that my description of the school "must somehow acknowledge the differences within and among the stories of experience, how they are told, and what it is that structures the telling and the retelling."

Instead of looking for pieces to the puzzle that I called ESH, I came to see girls' partial and competing stories as rhetorical performances of their school. In calling girls' stories of ESH rhetorical performances, I aim to draw attention to them as persuasive acts on three different levels. First, each story represents a girl's desire to authenticate her own understanding of what was going on in the school. In telling stories about ESH, girls persuaded themselves that their views made sense and had meaning in relation to their own lived experiences. Second, each story represents a girl's desire to persuade me of what was going on in the school. Often, a girl would begin her story by saying, "Everyone thinks the school is like *this*, but really, it's like *this*," or "I don't know why people think the school is like that, because it isn't!" or "Let me tell you what's *really* going on." Ling and Ana wanted me to know that ESH was not what I might have thought based on representations of high schools in Hollywood films. These remarks suggest that girls enjoyed giving me the "scoop" on their school as key informants who rightly felt that they were offering me important and useful information. This second form of persuasion created a sense of authority and influence for the teller, who was able to assert her opinion of the school with the confidence of an experienced insider. Finally, on a third level, I hope to "persuade readers of the credibility of my interpretive efforts" (Britzman 2000, 33) here, in this chapter. My own rhetorical performance of ESH has been influenced not only by girls' persuasive accounts, but also by my own understanding of what was going on while I was in the school. While the two are inherently connected, I gleaned my own socially situated perspective of ESH based on the classes I attended and the girls I chose to hang out with, all of which shaped my view of the school.

By callings these stories rhetorical performances, I also aim to draw attention to the performative aspects inherent in the making of a school.[4] A school is made by the bricks and walls that give it physical shape, by the rules and regulations that give it social structure, by the bodies that give it purpose, and by the internal and external discourses that institutionalize it as that thing we call "school." Schools are not transhistorical sites that exist outside of these multiple constituting mechanisms. As Butler (1990, 1993) notes in relation to gender, bodies are continuously constituted by practices that inscribe them as male or female, normal or abject. Through this inscription, bodies become recognizable and thus sanctioned within the social world; they become known, easily understood, and familiar. Butler (1990, 33) thus defines gender as "a set of repeated acts within a highly regulatory frame that congeals over time to produce the appearance of a substance, of a natural sort of being." Rhetorical performances suggest a similar feature of schools; schools are produced and perpetuated through constant and multiple performances that inscribe them as that thing we call "school." For Foucault (1972), discourse's most innovative effect, then, is not what *is* said, but what gets repeated and circulated, congealing over time and space to become "truth."

A school is performed by insiders: students, teachers, staff, and administrators, but it is also performed by those outside of its walls, extending its institutionalization into the neighborhood and the city in which the school exists. These external discourses are based on what people think is going on in the school from an outsider perspective, including rumors, media reports, and swirling accounts about the "kind" of students who go to that "kind" of school. Such internal and external discourses impacted girls' rhetorical performances of ESH. While girls had a sense of the school based on their socially and culturally situated experiences, external understandings that circulated throughout the city of Vancouver and the neighborhood of Wellington mediated that opinion. Yon (2000, 32) refers to this set of competing discourses as "the discursive space of schooling," where "various fragments of discourses" are juxtaposed "to consider how these act upon the actor's view of what is going on." This chapter is an engagement with girls' understandings of what was going on at ESH based on external discourses that circulated throughout the city and neighborhood, as well as internal discourses that worked to counterbalance the negative impressions that the school seemed to generate outside of its walls. Finally, I focus on the rhetorical performances girls offered of their individual school programs, specifically the ones from which I drew my sample: the Regular program, the Aesthetician program, and the French Immersion program. In repeating the stories that girls told me about their school, I inevitably become a part of its institutionalization, contributing to the circulation of internal and external accounts. But in framing these stories as rhetorical performances, my aim is to give readers a sense of ESH without reducing it to a false unity, and to highlight the ways in which both schools and school identities are "made."

"It's Our Reputation"

ESH is situated on the east side of Vancouver, a metropolitan Canadian city with a population of just under two million people. Vancouver has many reputations that revolve around its Pacific northwest location—hippy city, ritzy city, outdoorsy city, pot-smoking city, rainy city, granola city, lotusland, and the yoga capital of Canada. While often voted one of the best cities to live in the world, Vancouver is also home to what is commonly referred to as the "poorest zip code in Canada." Aside from being oceanside, mountainous, and host to the 2010 Olympics, the city also battles an overwhelming heroin problem in its downtown east side and is home to North America's first safe injection site for heroin addicts. Amidst these competing reputations, Vancouver also has the highest percentage of visible minorities in Canada.[5] One in three Vancouver residents is Asian and half of that percentage is Chinese.[6] These statistics have earned Vancouver yet another reputation—the city with the largest Chinese population outside of China. Vancouver also has high percentages of East Indians, Punjabis and Pakistanis, Filipinos, Koreans, Japanese, and Vietnamese. Its substantial immigrant population further emphasizes the cultural diversity of the city. Roughly 40 percent of Vancouverites are born outside of Canada, in places such as China, Hong Kong, Vietnam, India, and the Philippines. This percentage surpasses the immigrant population of almost any city in the world.[7]

Within this global environment, Vancouver has not become a city of segregated ethnic neighborhoods. It is not that Vancouver has no ethnically infused communities; on the contrary, such an infusion is unmistakable throughout the city. But neighborhoods are not the specific domains of any one cultural group.[8] The steady arrival of new immigrants and their dispersal throughout the city has created a ubiquitous ethnicity, but one that is not wholly recognizable in so far as it is contained within a geographic location. New immigrants from India might settle in Chinatown, for example, just as Chinese immigrants might settle in Little India. As Paul Delaney (1994, 8) writes, this ethnic diffusion goes beyond multiculturalism or the "modernist notion of immigration." Rather, he calls it "postmodern transculturalism and hybridization." In the postmodern city of Vancouver, "we all cross and re-cross ethnic borders everyday." For Delaney, this postmodern milieu means that we can only speak of Vancouver "using metaphors like the mosaic, the network, the collage, or the marketplace" (5). No other way of describing the city seems to suffice.

The metaphor of the marketplace is a fitting one for Wellington, the east side neighborhood in which ESH is firmly situated and where most students at ESH lived, worked, and hung out. Wellington Ave., the street that embodied the *esprit de corps* of the neighborhood, is known as Vancouver's funkiest area, attracting university students to its coffee house cool, artists to its lack of pretension, and hip urbanites to its raw and edgy reputation. The promise of "street credibility" coupled with the working-class and immigrant populations that occupy the area have formed a marketplace state of mind on the street, where people mingle from a wide variety of racial, ethnic, and economic backgrounds. As a result, Wellington is perceived as a fluid community of diversity, offering its eclectic residents the chance to socialize on any given day at one of Wellington Ave.'s Italian gelato parlors, Moroccan cafés, Ethiopian eateries, urban juice bars, Portuguese bakeries, Chinese restaurants, and Lebanese falafel joints. To walk down Wellington Ave. is to experience exciting juxtapositions of culture, such as old Italian men drinking espresso, punks shouting out to passersby, Chinese couples shopping for produce, performance artists engaging in their craft, street youth panhandling, and young urban professionals awaiting the next commuter train downtown.

ESH was a product of this diverse and dynamic environment, encapsulating the racial, ethnic, and economic multiplicity of the neighborhood. But the school was burdened with a "bad" reputation in the city due to its geographic location, immigrant, working-class, and working-poor populations, and government designation as an "inner city" school. When I mentioned that I was conducting research at ESH, people often furrowed their brows and wished me "luck" in such a tough environment. Students and teachers at ESH were well aware of this reputation. "Everyone outside of ESH thinks it's a scary place," Ms. Mackenzie told me. She was one of the youngest teachers on staff and had a genuine interest in research on girls. While most teachers were polite and tolerant of my presence in their classrooms, Ms. Mackenzie was an ally who helped me to navigate my way around the school. She told me that people often had a "strong" reaction to her job as a teacher at ESH and associated her school with getting "beaten up," "gangs," and "violent crime." "They think everyone at ESH is in a gang," she said with some frustration. "It's our reputation." She explained that the media painted ESH as a "bad place" and that this negative view had been "engrained" in people's minds. "Any crime in the city must be related to Vietnamese gangs, or any car racing must be Asian youth driving. You know, it goes on and on. And that happens to be *our* population, so we're forever getting slammed for it in the paper." Even the principal had been affected by ESH's status as a "scary" school. Several students had complained to Ms. Mackenzie that the principal was treating them as if they were already "in trouble." As one student noted, "He treats us like we're not a good school, and he doesn't know how good we are!" Some of the students had hoped to sit down with the principal and let him know that "ESH is not as bad as he thinks."

"EAST SIDE PRIDE"

ESH was a school lacking in luxuries. While it had everything it seemed to need, it did not have everything it might have wanted. Punctuated with visible pipes along the ceiling, a gym from the 1950s, and bath-rooms that still had the high "flush" levers of another era, it was cer-tainly not modern. It had been cobbled together during various decades in the past that betrayed the school's lack of funds for finishing (or starting) projects—a discernible difference from the construction job that was underway at Beach View, an affluent west side school, where floor-to-ceiling windows, an outdoor basketball court, and a landscaped picnic lunch area had been constructed during its most recent renovation project. ESH had no such extravagances and girls were aware that their school was "not rich." In Industrial Arts class one day, I was sitting with a group of girls who made their knowledge of ESH's financial troubles clear: Francesca, a deadly sarcastic Italian girl from an immigrant fam-ily, Cookie, an Aboriginal girl who could always be counted on to laugh at Francesca's disgusting noises and dirty jokes, and Wanda, an African Canadian girl who pumped me for information on university life when-ever I saw her in the hallway. During the class, the girls and I chatted about their school while they polished their recently forged metal rings. Francesca wondered if I knew that ESH was on "welfare," because it re-ceived "lots" of money from the government to "stay afloat." But before I could respond, Francesca yelled, "I don't care if it *is* on welfare, it's where it's *at!*" This declaration elicited a round of whoops and hollers from Cookie and Wanda reminiscent of the *Jerry Springer Show*. Then Cookie loudly added, "ESH is *the* place to be!" At that point, many other stu-dents turned to look at our corner of the table, cheering under their breath so as not to arouse the suspicion of Mr. Humphries, the strict Industrial Arts teacher.

This kind of school pride was not hard to find at ESH, even amidst a general dislike for classes, teachers, and the administration. An underdog spirit born of ESH's "bad" reputation helped to rally students. The girls' bathrooms were filled with "East Side Pride" sloganeering and there was a well-known hand gesture that students flashed to each other in the halls as a show of solidarity: Three fingers on the right hand turned side-ways to create an "E." In fact, it was hard to get through a day at ESH without hearing someone extol the virtues of being from the east side. As Yon (2000, 33) found during his study at an inner city Toronto high school, "negative associations produce a sense of cohesiveness, mutual dependence, and community because of a 'pervasive' feeling of being threatened." ESH was similarly unified as the students were constantly facing citywide stereotypes about their school and what it meant to live on the east side. But this "bad" reputation fostered a "generally positive" attitude inside the school based on "a communal underdog spirit result-ing from the school's negative image" (Bettis 1996, 116).

East side and west side Vancouver have well-established reputations throughout the city. Though both geographic areas are heterogeneous and have large immigrant populations, the economic disparity between them has created reputations that are as well known as the street that separates them geographically.[9] The west side is known to be "upscale," "yuppie," and "trendy," inhabited by white and Asian professionals and their families. Though the west side sees crime, homeless, and drug related activities, these issues are often quickly dealt with by urban security guards and responsive police departments, who take extra care in keeping west side neighborhoods secure. As Abby whispered to me on a class trip to see *To Kill A Mockingbird* at a west side theater, "It feels so *safe* here." Within this sanitized reputation, west side schools are viewed as academic, brimming with resources, and populated by rich white and Asian students who are university bound. As well, west side parents are seen as caring, dedicated, and generous in their desire to help out both inside and outside of the school.

Conversely, the east side is known to be community-oriented and down-to-earth, characteristics that intersect with its reputation for poverty, crime, gangs, and drugs. Though middle-class homeowners have flocked to the area to enjoy affordable mortgages, the reputation of the east side remains predominantly working-class and immigrant. Adding to this reputation, the east side sees more illegal activity, homelessness, and visible drug use than the west side, reputations that fix it in the public imaginary as "dangerous."[10] Within this negative view, east side schools are generally viewed as vocationally-oriented, run down, lacking in resources, academically challenged, and populated by poor immigrants, Aboriginal students, and gangs who deal drugs.

This view of the east side—held by many (but not all) outside the area—is further emphasized for girls. East side girls are seen as "cheap," "trashy," and "skanky" by some students in west side schools.[11] Conversely, west side girls are seen as "snobby," "cliquey," and "spoiled." Keisha, a working-poor African Canadian girl in the Aesthetician program, noted that while ESH had some "snobby elements," it really "wasn't *that* bad compared to the schools on the west side." As Keisha explained, in west side schools you were judged very "harshly" for what you did, or did not, wear.

Yeah, like if I went to Beach View and had the clothes I have? And I didn't dress in *Miss Sixty's* and I didn't have all the stuff from *Off the Wall* and *Mavi* and all that,[12] then they'll look at me like, "Oh, she's poor, I don't think she has money. Why's she going here? We're west side kids. We're rich. We're supposed to wear the *Miss Sixty's* and everything like that. You're supposed to look high class!"

In agreement with Keisha, Mallory, a girl who had just transferred to ESH, stopped to chat with me in the hall outside of French class one day. "How do you like it here?" I asked. "So much better," she sighed. "At Braeburn [another affluent school on the west side], the girls are so mean. Here, they're way nicer."

ESH's internal reputation as a "nice" place filtered through many of the conversations I had with girls. This view seemed to lessen the wounds some experienced through their discursive positioning as "east side girls." But many girls also took on the east side identity as a badge of honor. To most girls, coming from the east side meant that they had seen more, done more, and knew more than their "sheltered" and "pampered" west side counterparts. Shelley, Stacey, and Marla displayed their street smarts during an assignment in English class one day. They were asked to draw maps of their childhood homes. With a good deal of laughter, the girls began sketching out the crack houses, prostitutes, heroin needles, and mugging incidents that they felt were a part of their neighborhood. Living in Wellington offered girls these and other lessons in life. For example, two drunken men had recently propositioned Mina as she walked home from school one day. As she explained, "I've had offers to—you know on Wellington Ave. they have places that have open bars, like *Frisco's* or *Café Ouzo?* I've been offered by a couple of guys to maybe have a couple of drinks and go back to their place." Because this text does not allow for Mina's tone to come through, I will add that she told me this story—and her ability to "blow them off"—with extreme pride and pure satisfaction.

"Our School Lets Us Be Diverse"

While the external reputation of the school constructed the east side as a harsh place, an entirely different story bubbled up from within. Its student body saw ESH as a "laid back" school with an "anything goes" atmosphere. Students also viewed it as a "safe place." As Azmera said: "I like school. I like school because there's a lot of different people. And I don't know. I just—I like the environment and for me school's a safe place." ESH was a community school in that its reach extended into the Wellington neighborhood. It offered adults art classes, computer seminars, language courses, and spaces for meetings during off-hours. It was a place for older Chinese men to practice Tai Chi at the crack of dawn, a place for teenagers to hang out and play basketball, and a place for students to get a free hot lunch if they needed it. As a result, the school was much appreciated within the Wellington community. It acted as a focal point and was a hub of social activity and a hangout. As some teachers explained to me, the school was a place for students to "just be" if there was "trouble at home."

This positive internal reputation was also predicated on what girls named as the "best thing" about ESH: diversity. Diversity was defined by almost everyone as not only multiculturalism but as also social and curricular variety. For Chrissie, ESH's extreme range of social scenes, groups, and cliques was easily juxtaposed to the "conformity" that she witnessed at her old west side school. "There's always this concept of like, popular cliques and stuff like that. And at ESH, it's like, it's really not like, there are different groups, but I can't think of like, a popular group. At Braeburn it's different." Braeburn was where Chrissie had gone for a few months before switching back to ESH.

> At Braeburn, you can see the defined little groups of kids in the hallways. Like, they stand in circles and like, if you look around at my social group at ESH, Dren and Ratch are totally punk, and then some of my friends are dressed really conventionally and stuff like that. And then, at Braeburn, all the girls who dress one way stand in this group, and all the like, goth/punk people stand in one group, and you can see them all dressed in black, standing in one little circle.

When I asked if ESH was even just a little bit like that, she responded, "ESH is *not* like Braeburn at all! At Braeburn there's the cheerleaders, there's the jocks, there's the goths, and you can see them all. They all hang out in clumps." But according to Chrissie, at ESH, the social groups were much more eclectic and open to difference.

Others agreed with Chrissie's construction of ESH as socially diverse and this rhetorical performance seemed to perpetuate the idea that ESH was an egalitarian school, where, as Ling and Ana said, "all groups are even." Zeni noted that she did not think hierarchies existed at ESH: "There's not really 'groups.' I think everybody gets along well. I think we're all the same." Azmera, too, told me that she thought, "everyone is pretty much, is pretty equal." Dren indicated that popular "people" existed, but not as a readily identified group. "There are a couple of people everyone wants to be friends with and blah, blah, blah. But a lot of other high schools have that one big group." She went on to explain, "Every class of twenty had a couple of popular people. But if you asked me straight-out, 'Who's the popular people?' I would not have the answer."

As I noted earlier, not all girls saw ESH as an egalitarian school, but there was a strong desire among many of the girls I spoke with to construct ESH as devoid of any hierarchical structure, or as Ling and Ana suggested, as a school that was "not like the movies." While there were hierarchies present at ESH, they were indeed different from the central social hierarchies so often depicted in Hollywood representations of high school life. In such representations, mutually defining groups coexist in binary opposition to each other, organized around a single identity category, such as class or race. But at ESH, a central social hierarchy gave way to "multiple social hierarchies" (Bettie 2003, 5). No one hierarchical structure dominated the school, where an easily identifiable group of popular students sat at the top and the other social groups followed in a linear fashion down to the bottom. Such a straightforward hierarchical structure would have been impossible to support at a school brimming with so much cultural diversity. Hierarchies existed among particular racially defined groups, such as the "hardcore" or popular Chinese students, the "average" or slightly less popular Chinese students, and the "quiet" or least popular Chinese students. Hierarchies existed among Canadian-born cultural groups and groups who came from outside of Canada. Hierarchies existed among academic and vocational programs, among the various Asian groups, and among the preppy girls and those who classified themselves as alternative, sporty, or dressy. But these hierarchies did not add up to one overarching social pattern that sustained itself across the school. The school's large population of 1,700 students and its system of multiple curricular tracks made it very difficult for anyone to really know who was supposed to be "on top" at all times. As a result, "top" and "bottom" continuously shifted depending on whom I asked.

ESH's multiple social hierarchies were also predicated on its global milieu. In the school, its culturally diverse student body spoke over fifty-four home languages. Though many urban schools across North America are shaped by these "global times" (Yon 2000), school ethnographies still tend to privilege a central social hierarchy supported by a singular orientation to identity, such as race, gender, or class. Such an example can be found in Penelope Eckert's (1989) classic *Jocks and Burnouts*. In Eckert's study, Jocks and Burnouts are representative categories that exist in mutual definition. Jocks are characterized as middle-class students who are college-bound, athletic, involved in school culture, and "party" only on weekends. At ESH, such a singular orientation to identity was contradicted by many of ESH's female sports stars who were working-class, vocational students. Keisha, a student in the Aesthetician program who planned to start full-time work just as soon as she graduated, switched to ESH "for the sports." As well, many of the student council representatives—positions occupied by "Jocks" in Eckert's study—were working-class girls enrolled in one of ESH's advanced academic programs and thus working toward university admittance. In short, the "cooperative" relationship that Eckert shows "Jocks" to have with the school was not reserved *just* for middle-class students at ESH.

Similarly, the category "Burnout," which Eckert defines as someone who is working-class, enrolled mainly in noncollege-prep courses, geared toward a vocational occupation, and "parties" all week—particularly with harder substances—could have applied in some manner to many middle-class students at ESH. The skaters and "skater affiliates" (Kelly et al. 2005) often described themselves as "stoners," "slackers," and "pot-heads." But skaters were largely middle-class students enrolled in advanced academic programs, features that did not coincide with Eckert's recognition of an adversarial relationship between Burnouts and the school. Similarly, many of the popular Chinese girls, known as "hard-cores," had reputations for doing drugs, drinking, swearing, and "blow-ing off" school. Yet some of them also came from middle-class families and were enrolled in academic programs. Further deviating from Eckert's description of "Burnout," many working-class girls at ESH were new to Canada or first generation Canadians who wanted to please their immigrant parents by getting good grades and obtaining high status jobs in their new home country. They thus worked hard in school and strove for college or university entrance.

Eckert's study was done in a homogeneous Detroit high school during the 1980s and thus makes for a good—though somewhat unfair—contrast to the multiple social hierarchies of ESH. But it also helps to show how other schools are "made" by school ethnographers, who choose which features to highlight and which stories to tell. Each school is produced within different internal and external discourses, contributing to unique rhetorical performances by its students. At ESH, the global milieu had a profound impact on girls' understandings of their school. No matter which cultural location a girl occupied, she never failed to mention the racial and ethnic diversity of ESH as one of its "best things." This narrative may have been influenced by the fact that west side schools did not boast student bodies that were *as* culturally diverse as those on the east side. West side schools were indeed culturally diverse, but it was well known that east side schools contained a much wider expanse of cultures and home languages. This "added bonus" gave girls at ESH one more thing to flaunt about their school, particularly in the face of negative east side constructions.

Aside from cultural diversity, curricular diversity also came up as one of girls' favorite things about their school. ESH had numerous curricular tracks—academic, vocational, cultural, alternative, and special needs—that made it very appealing to a broad spectrum of students and parents. Jamie commented on how such a wide variety of choices made it "easy to be you." West side schools, she told me, were designed to "keep diverse people OUT purposefully!" But ESH is "very open with our classes." Jamie, who sported eight piercings and was often singled out as one of the "most pierced" people in the school, felt that the school's inclusive organization diminished the "risk" associated with being "different."

The majority of ESH's students were enrolled in the school's main student body, or what was commonly referred to as the "Regular" program because it did not claim to have a specific agenda, other than "regular" education. Outside of the Regular program, smaller populations participated in programs for ESL and learning disabilities. Numerous programs also existed that catered to advanced academics, vocations, alternative education (dropout prevention), and culturally specific learning (the Aboriginal program). ESH's academic programs comprised the "big three" in power and prestige: French Immersion, Accelerated Studies, and Independent Learners. The "big three" were populated by students from middle-, working-class, and working-poor backgrounds, but French Immersion tended to be mainly white, while Accelerated Studies and Independent Learners tended to be mainly Asian. ESH's vocational programs, ranging from the Aesthetician program to mechanics to carpentry to cook training, garnered less power and prestige. These streams gave students practical instruction in trades while they worked toward completing required high school credits for graduation. Most of the students in ESH's vocational programs were working-class and working-poor. ESH also offered programs based on the cultural diversity of its student population, specifically those from Aboriginal backgrounds. The Aboriginal program did not include all Aboriginal students within the school, however, but catered to those who needed or wanted the extra attention that a specialized classroom could provide. It was often viewed as a "dumping ground" for "troubled Native kids."

Though each program was housed within ESH, they all had their own space in the school, with designated teaching staff, and program identities and reputations. This organization meant that a girl could be popular within her program and virtually unknown within the school at-large. Based on the relative cohesion and isolation of each program, a girl's friends were most often, though not always, determined by her curricular track. As a result, each program had its own unique social structure. As Jamie explained, the programs "don't really mix themselves a whole lot. I don't have a lot of classes with the French Immersion people; English is my only one. So I make friends with the Regular students, because those are who I have classes with." Though this was not a hard and fast rule, it was commonly acknowledged that the smaller programs constituted "posses" who stuck together. There was a certain amount of cross-fertilization between academic programs and the Regular program in classes that serviced both, but as Ratch put it, "you would never get someone from drop-out prevention and someone from Accelerated Studies like, hanging out." Dren, Ratch's constant companion, emphatically added, "No! NEVER! Ever, ever, ever!" Such segregations made it easy for multiple social hierarchies to flourish within individual programs that acted as the basis for girls' most impassioned and detailed rhetorical performances. Below, I highlight some of the rhetorical performances girls offered in relation to the three programs from which I drew my sample: the Regular program, the Aesthetician program, and the French Immersion Program.

"IT'S LIKE REALLY ASIAN": THE REGULAR PROGRAM

If the Regular program had any reputation at all, it was for being Asian—a blanket statement that was used to describe the entirety of its largely working-class population. This vague description included a variety of social groups or "Asian sub-groups," as they were called, including Nammers (Vietnamese), Hongers (Hong Kong), Flips (Filipino), Japanese/Koreans (often viewed as one group), and Chinks (Chinese).[13] Like other Asian subgroups, the Chinese subgroup was further divided into categories that related to the length of time one had spent in Canada: CBCs (Canadian-Born Chinese) and ESLs (English-as-second-language).[14] Xiu, a first generation Chinese Canadian who called herself a CBC, told me that each group thought it was better than the other: "Yeah. It's um, I guess the CBCs will think, you know, 'I'm better than you [ESLs] cause I know more English and I have more experience here.' And then they're [ESLs] going, 'Oh you guys are so dumb. In Hong Kong I learned this already.' Yeah, it's competitive."

Karen Pyke and Tran Dang (2003, 148) refer to such "sub-ethnic identities" as forms of "intraethnic othering," or how people who occupy the same ethnic designation create forms of disidentification among themselves. These disidentifications often result in internalized racism, usually directed at the "more ethnically-identified, by other co-ethnics, usually the more assimilated" (152). At ESH, intraethnic othering worked in both directions. ESLs felt that they were better than CBCs because of their perceived cultural "authenticity," their loyalty to the home country, and their detailed knowledge of a language and culture that was not Canadian. And CBCs felt that they were better than ESLs because they had adapted to Canadian culture, were more interested in the present as opposed to the perceived obsession that ESLs had with the past, and were more aware of North American popular culture, such as style and music.

The Asian subgroups at ESH were largely organized around intraethnic othering. Yet further distinctions existed among them that Xiu categorized as "hardcore," "average," and "quiet," denoting various levels of popularity. She explained these categories using her own Chinese social scene: "hardcore Chinese" were popular, tough, fashionable, and swore "a lot," "average Chinese" were somewhat popular and played cards and computer games at lunch, and "quiet Chinese" were not that popular and studied constantly. A few other Asian subgroups were mentioned in conversations I had with white girls (who were called "the Whites" among Asian circles), including the "straight-edge" Chinese girls, often associated with Christianity and clean living, and the Asian computer "geeks," "nerds," or "dorks," who played cards by the library and were often associated with ESL.

As Xiu made clear, Asian hardcores, specifically Vietnamese and Chinese girls, were "very popular" within the Regular program. As a justification for this assertion, she noted that, "usually in classrooms, you'll mainly see black hair students." This majority fostered an assumption among other students at ESH that Asian girls were one "big" group who were all the same. Teachers, often unaware of most of the Asian subgroups, viewed their classes as relatively uniform. Asian girls were sometimes characterized in a lump with words like "nice," "quiet," and "smart." Mr. Murphy, an English teacher, told me that he only had "very well-behaved" Asian girls in his classes. Non-Asian girls often expressed similar feelings. Chrissie, who was Jewish, mentioned that "the Asian girls are pretty close, *all* of them. And they hang out together a lot." She theorized this explanation: "Possibly with the Asian people, they have that same background or something, so they have something in common right off. And maybe that attracts them to each other." This perceived homogeneity among Asian girls produced some resentment among the non-Asian students. Dren explained that, because "white people are definitely the minority at my school, I think that they do tend to, see this is a lot of generalization here, but they do tend to socialize a lot more with the other races than the, like, Asian people."

But this supposed homogeneity overlooked the myriad segregations that existed among Asian subgroups. Zeni, a Japanese and Filipino girl, told me that Chinese people don't really hang out with Japanese people" based on "the ancient days" when "the Japanese went over to China, you know, and tried to kill everyone there." She went on to explain that the Japanese also "tried to take over the Philippines too. A lot of people don't like the Japanese." Xiu mentioned the antagonism between the Chinese and those from Hong Kong. "I don't know how they get divided like that," she initially wondered, "cause Hong Kong and China are like one." But she quickly amended this statement: "Well, they're one big place, but they're still separated, cause then the Hong Kong people will say, 'Oh, the Chinese people are stupid, or they're poor or we're better than them!'" She similarly described a dislike between the Chinese and the Filipinos. When a Filipino boy asked Xiu out, she told me that she "had to say no," because Filipinos were not "suitable" boyfriends according to her family.

While Asian girls were often viewed as exclusionary, they were not the only racially segregated social scenes in the Regular program. The Hispanic girls also had a reputation for hanging out on their own. Zeni explained that Hispanic girls "all hang out together. Doesn't matter what grade they're in. As long as they're Spanish." When Zeni started at ESH, the Spanish girls found out that she was half Filipina and approached her to be in their social group. As she put it, "as soon as I went to the school, they asked me what race I am. And it's kind of like, 'I'm half Filipino.' If I'm Spanish, the Spanish girls are just going to come up to me and kind of like accept me into their group. And it was kind of like that." She explained that she was "welcomed to the group" as "kind of Spanish-y, Portuguese."

But the Regular program was also full of countless mixed racial groups. Isabel, who was Hispanic, hung out with girls who were Persian, Cambodian, Portuguese, and African Canadian. Gwen, who was Aboriginal, told me that her group of friends included "Asian kids, Native kids, white kids." And Sydney, who was Japanese, white, and Hispanic, hung out with girls who were Italian, Aboriginal, and African Canadian. Mixed race and racially segregated groups coexisted within the Regular program without much overt racial tension; however, there was one well-known form of racism in the school that was directed at Aboriginal students. During a rather difficult conversation, Gwen quietly told me that if you were Native at ESH, "people sort of look down on you. It'd just be like, 'Oh, she's Native. She's stupid.'" As a result of this negative perception, Gwen chose to work very hard in school in order to avoid the "lazy" and "crazy" Native stereotypes that were perpetuated at ESH. "Yeah, I think I work really hard about it," she told me. Gwen felt her Coast Salish background, a First Nations group indigenous to the Vancouver area prior to European colonization, marked her as "different" from all the white and Asian girls "upstairs." "Upstairs" and "downstairs" were terms students used to describe various program locations. Gwen spent alternate days "downstairs" with students in the Aboriginal program and "upstairs" with students in the Regular program. As Dren explained, "if there's racism at ESH, it's against the Native kids." Ratch concurred: "They're, like, treated like the lowest of the lows." But according to Dren and Ratch, it was not the white students who were "racist against Natives." They told me that Asian students were "the really racist ones" who called Aboriginal students "chugs." "Chug" was a derogatory label used by Nammers and other Asian subgroups—a word that was short for "chugging," meaning to drink alcohol quickly. Though Dren and Ratch were quick to point the finger elsewhere, I constantly observed racist sentiments directed toward Aboriginal students by white students as well.

These reputations permeated the Regular program and gave it both a pluralistic and antagonistic feel. Because the program included most of the students at ESH, almost everyone had an opinion about what was going on in the Regular program, where girls' rhetorical performances were bound up in racial and ethnic positionings, as well as racist tensions and intraethnic othering. While mixed race groups were common, and racial and ethnic groups coexisted without overt racial tensions, smaller-scale hierarchies predicated on race, ethnicity, and nationhood created status systems. But not all programs at ESH were similarly experienced—especially not the ultra-pluralistic Aesthetician program.

"Probably All Blonde": The Aesthetician Program

The Aesthetician program, aka beauty school, was a vocational stream that operated in a wing that many students at ESH never visited. It was in the "Spanish" hall, so-named for the high concentration of Hispanic girls who hung out there, only some of whom were in beauty school. But like the Regular program, beauty school was by no means homogeneous. It attracted girls from many cultural locations, creating one of the most racially diverse programs in the school. As Gianna, an Italian beauty girl, told me,

> I don't even think there's anything to do with different races [in our program]. Like, everyone just hangs—like, if you see us, we're all different. We have friends that like—look! Azmera is like, you know, black. We've got black friends, Spanish friends, Chilean. We've got Asian. We've got, you know, Vietnamese, Chinese. Everyone's just mixed. And even within us, like, like Punita. She's like half brown, half Italian. Like, everybody's just all mixed and we all get along.

Beauty school was made up of a small group of students who traveled together throughout most of the day. And all of the students were girls.

Keisha mentioned that they would like to have boys in the program, "just to see how it would feel," but the "all-girls" atmosphere was "kind of better cause then we can talk about anything!" And they *did* talk about anything, from boys to sex to birth control to cramps to makeup. Ms. War-chowski, their good-natured teacher, let the girls listen to the radio while they performed practical tasks on each other, creating an environment where singing, dancing, swearing, and gossiping were commonplace. Shen explained that in beauty school, girls were allowed to "speak their mind without teachers telling you to be quiet," a quality she relished. Isabel also noted that being with the same girls "24/7" created a sorority setting based on "feminine" issues. "Like, we can talk about any girl stuff, and we don't have to worry about a guy, like, being there. Like, last class, we were wax-ing our armpits and no one has to care, you know?"

The "family" mentality that most agreed was fostered within the pro-gram spilled into lunch, when the beauty girls would emerge from the "Spanish" hall and commandeer a table in the cafeteria. They were the loudest table in the room and they knew it. They delighted in drawing attention to themselves, stridently invoking their neighborhood savvy by discussing their Saturday nights on Wellington Ave., their work experi-ences around the city, and their dates with men (as opposed to high school boys). While working-class students were in the majority at ESH and, therefore, a majority in most programs, the girls in beauty school seemed to be the most conscious of their working-class positioning. As Isabel noted in relation to her own background, "I think we're a working-class family. We're not rich, nothing. But we've been really blessed. Because of my parents, how much they've worked. Because I've seen them work. And it's hard."

Many of the girls worked at salons or spas across the city as part of their certification requirements and this "real world" experience enabled them to comment on their "well-to-do" west side customers. Gianna worked at a ritzy salon on the west side that catered to "older women." She described them as having "lots of money, you know? Some of them never worked in their life!" Because apprenticeship was such an integral component to the beauty school program, it fostered a connection be-tween education and work that other girls in the school did not often experience. Work was *real* to beauty girls. They were required to obtain and keep jobs in the trade, but they also depended upon work in order to contribute at home and still have enough left over to buy their own clothes.

Gianna was well aware of the connection between school and work, and this awareness kept her in the Aesthetician program, even though she had dropped out several times.

And then I just decided that, I'm like, "Obviously you have to finish school. Where are you going to go if you don't finish school?" And I always say to people, like, that's my thing, like, if you're not educated, like, I can't stand when people are talking and they just like, they're so uneducated! Or they just talk and they don't even know what they're talking about. So then I was like, "Fine! I'm just going to go back and finish." Like, for my mom too, right?

Gianna believed that education could take her anywhere she wanted to go and she proved this point by comparing the level of education achieved by her friends on the east side to those on the west side. "Just go down to the Projects and knock on their door and ask them how many people have graduated," she said emphatically, "and go to the west side and knock on their door, right? They'll be like, 'Oh, yeah! Graduated and took five years of university!' Education is like, the *key*."

The beauty girls shared many of these working-class experiences and dispositions, deepening the "family" mentality among them and engendering a powerful form of loyalty. Keisha explained that this loyalty was fostered by "being with each other all day. And you just learn to like adapt to one another, and then—you just stick with each other. You support each other! *We're the beauty girls,* you know?!" Aware of their outwardly visible solidarity, they described themselves as "powerful," "outspoken," "equal," and "the closest group in the school." Jo summed up the pluralism of the program: "Everyone accepts everyone for who they are, you know?" Ironically, Jo was one of the few girls who occasionally took some "gentle" ribbing in the program for being "slow." If someone asked for clarification or extra time, everyone assumed that it was Jo. But Jo certainly never complained about being teased, and Gianna softened the effects of the banter by ensuring that, "we're just doing it for fun. We don't want to hurt anybody's feelings."

Most of the beauty girls smoked, and they all took their smoke breaks together, tumbling out of the classroom in a pack while laughing over a juicy piece of gossip or the visibility of a girl's thong in her low-rise jeans. They ate their lunch together, went to dances together, and watched each other's backs. When they entered the cafeteria as a gang, Isabel felt that people recognized them as the "most beautiful girls in the school." But the beauty girls were not seen in this "flattering" light by everyone, and most of the girls in the program knew that they had garnered other reputations for being "loud," "dumb," "bimbos," "airheads," "gossipy," and "lazy." As Azmera overheard a girl in one of her Regular program classes say, "oh, the reason why they take beauty school is probably because they're so stupid they can't take any other classes!" Outraged,

Azmera shouted, "*I'm* in beauty school and I'm *not* stupid!" The girl responded, "'Yeah, well whatever. All the girls that are there are probably *all* blonde.'" This last remark, Azmera explained, was meant to indicate that beauty school is "such a girly thing and we're not capable of anything else."

Shen experienced similar negativity, explaining that most people in the school thought that girls only went into beauty school if they "don't do well in school academically" and if they are "*all* about looks," a reputation that was enhanced by Ms. Warchowski's rules of "professional" dress for the program. The beauty girls had to wear "industry" type outfits and makeup (especially lipstick) during their beauty school training, particularly when clients came in during "spa days." This did not mean business suits or white blouses; it meant glamorous styles that were often perceived as "skanky" and "over-the-top" by students in other programs. This look earned many beauty girls the reputation of "hoochie," a racially designated term generally reserved for Hispanic and Aboriginal girls but also applied to girls of any cultural background who wore high heels, "hot" pants, "sexy" tops and had thin, penciled eyebrows, and tight, pulled back hair.

The beauty girls thus ranked fairly low on the school's curricular hierarchy, but did not seem to care. They viewed their vocational classes as the best part of the day and credited the program with keeping them in school, getting them a "decent" career, and instilling in them a sense of pride and accomplishment for a job well done. They were also well sheltered from the rest of the school and only really interacted with the Regular program students in their academic classes every other day. As Gianna admitted, "I don't generally know what goes on on that *side* of the school," indicating her lack of knowledge regarding advanced academic programs, such as French Immersion.

"THEY THINK THEY'RE ALL THAT": THE FRENCH IMMERSION PROGRAM

Ms. DiAngelo, who taught customer service class in the Aesthetician program, was quite plain about the difference between the beauty girls and the French Immersions. The Frenchies were "middle-class" and the beauty girls were "working-class." Because of this "obvious" disparity, she preferred to teach the working-class girls, as the middle-class ones did not "need" her as much. The parents of French Immersion students were already invested in their educations, and they were well supported both at home and in the school. But while Ms. DiAngelo's observations about the parents of French Immersion students were generally accurate, she overestimated the economic homogeneity of the program. Within French Immersion, at least one third came from working-class backgrounds. The fact that all the Frenchies were perceived to be middle-class was, perhaps, a function of their uncharacteristic racial homogeneity. Though not all white students at ESH were in French Immersion, almost all the French Immersions were white. The exceptions to this rule included small percentages of East Indian, Chinese, and Middle-Eastern students, all of whom hung out together. Shitar, a working-class Lebanese girl who was a part of this non-white contingent, told me that all the Frenchies "of color" ate lunch together everyday in the stairwell. "No whites with us." The other girls in French Immersion, whom she sarcastically called "the blondes," "don't have any Chinese friends. They don't have any black friends. They don't have Spanish friends. They're *all* white."

The majority of the grade ten Frenchies knew each other very well because they had been together "since kindergarten." From Shitar's point of view, this familiarity had turned them into a "dysfunctional family." But unlike the beauty girls' family, the Frenchie family did not welcome *everyone* in the program with open arms and instead operated on elements of exclusion. When I asked Shitar what role she had in the Frenchie family, she replied in a deadpan voice, "forgotten child." More than any other program, French Immersion seemed to embody a traditional social hierarchy that was likely fostered by its mainly white population. Within French Immersion, popularity revolved around three noticeable social groups: "loud" or "mean" popular girls, "quiet" or "nice" slightly less popular girls, and "weirdos," "outcasts," and "wannabes," all of whom existed on the lower rungs of the popularity ladder. This internal social hierarchy earned French Immersion girls the reputation for being "incredibly cliquey," "snobby," and "snots," particularly by those who found themselves at the bottom of the pecking order or by those from the Regular Program who encountered Frenchies in their classes.

The "loud" popular Frenchies earned their name by being confident, energetic, and assertive in and out of class. Leah, a "quiet" popular girl told me that this social group was "really kind of loud and really kind of in-your-face and, I don't really want to say the word obnoxious, but it's true. They're obnoxious." She explained that if you stood by their lockers and listened to their conversations, you would hear all about "hair, makeup, and the new person that everybody's decided they should like, or something." According to Diane, who existed on the "lower" rungs of the popularity ladder and initially worked very hard to become one of the "loud" popular girls, the features that enabled them to be "cool" were "the fact that they were all pretty. They all wear nice clothing. And they all, you know, talk about dancing and that kind of stuff." Abby, a self-defined "weirdo," described the girls at the top of the Frenchie hierarchy in similar terms. "They all shop at *Off the Wall* and have, like, MAC makeup. I don't know, they all listen to their stupid like, Nelly music."[15]

The characteristics used to describe the "loud" populars by other girls often fell under the rubric of "preppy," a word that had two meanings within the social world of ESH: an orientation toward schoolwork and a style. The "loud" populars fell into the latter category of preppy. They were mostly middle-class and wore "trendy" clothes, including low-rise jeans, midriff-revealing tops, and brand name skater shoes. They were generally thin and most of them, in keeping with Shitar's nickname, had long, blonde hair. While they certainly did not label their own style "preppy," it was a description that other girls easily applied to them. Diane described their style as "the tight jeans and the little shirts and the skater shoes and all that kind of stuff." She also explained that they were into "boys and clothes and magazines and music that was *really* bad." Zeni, who had English class with Frenchies, explained that you could tell the "loud" populars by the fact that they "have their shirt all the way up *here*, pants all the way down *here*." Zeni called them "very loud" and described them as having "their own little world" revolving around "guys."

Chrissie, a "loud" popular, saw herself as a "floater" who hung out with "lots of groups," such as the skaters. By association, the "loud" populars were also considered to be skaters, even though they did not skateboard. They saw themselves as part of skater culture through style, music, and their skater boyfriends. They thus described themselves as "laid back" and "mellow." They also saw themselves as "intellectual" and "deep," reputations that others sometimes agreed with. Dren, who started out in French Immersion but later switched to the Regular program, had mixed feelings about the "loud" populars, but agreed that they were smart. Dren put it like this: "I mean, I wouldn't necessarily say I agree with a lot of their views, but they all think about shit a lot and they all know a lot of stuff, and like, they all have really high standards." Their "intellectualism" was linked to creativity and was often juxtaposed to the "conformist" mentality of smart Asian girls.

The Frenchie reputation for "intellectualism" extended to the "quiet" popular girls as well, a social group that saw itself as quite different from their "stereotypically popular" counterparts. A stereotypical popular girl, according to Leah, was "really pretty, really loud," had "really nice clothes," and was "insanely obnoxious." In other words, "popular girls nobody really likes but everyone kind of wants to be." Conversely, Leah described herself as popular in the "other way," where "you have lots of friends and you feel kind of comfortable with everyone." Another marker of "quiet" popular status, according to Leah, was the ability to have "more interesting conversations." She and her friends often engaged in philosophical debates at lunch, earning them the reputation for being "artsy-fartsy" and "thinking too much." This reputation also earned them the title of preppy, but in both senses of the word. "Quiet" popular girls wore styles that were virtually identical to the "loud" populars, but also had a particular orientation to work; they not only did their work well, but also "enjoyed" doing it. As Mina put it, the "quiet" Frenchies "are people who just like, attack their homework with such tenacity that they're considered total preps, you know? They want to look good but they also want to get good grades." Mina, who was in Leah's English class, thought she was a quintessential prep. "Leah is one of those kids who will just get her work done as quickly and efficiently as possible and look forward to the next time she gets to do it."

SETTING THE STAGE FOR IDENTITY

The rhetorical performances girls offered contributed to the "making" of ESH as an institution, as a hang out, as a geographic space, as a social world, as a discursive construction, and as a school. Stories of pluralism and egalitarianism were juxtaposed with stories of social and curricular segregation. Stories of east side pride were juxtaposed with stories that circulated in the city regarding "skanky" east side girls. Internal discourses about the school being a "nice" and "safe" place were overlaid with external discourses of the geographic divide between east and west side Vancouver, and by ESH's reputation as a "bad" school. Stories about individual programs were juxtaposed with girls' social and cultural locations. This was the discursive space of ESH, a space that was constantly changing and open to myriad interpretations. Such a space offered limitless possibilities for identity construction and negotiation, given the room girls had to maneuver between social scenes, groups, and cliques. The sheer number of students in the school offered girls at least the *chance* of finding "their own people." But as open as the school seemed to be (to me, an outsider and adult), it was also clearly limiting in that girls were not always able to step back from their own social scenes to admire the vast expanse of ESH as a playground for identity possibilities. Instead, some girls felt trapped by their program reputations or by their restricted access to social groups. The postmodern milieu of the city had created the marketplace neighborhood of Wellington, which, in turn, produced the multiple social hierarchies of ESH. While the school was teeming with social possibilities that flowed within and across school programs, girls' everyday experiences often belied the freedom that such a milieu seemed to foster.

Girls' rhetorical performances of the school were intimately connected to their constructions and negotiations of identity. If identities are produced within constituting institutions, then the school's identity—however it may be conceived—is implicated in how girls saw themselves. Existing within a relationship of mutual definition and formation, girls and school contributed to an understanding of each other. The rhetorical performances that girls offered underscore the ways in which their identities were shaped by and continued to shape the school. The school was not only the constantly shifting stage upon which identities were performed, but also a constituting feature of identity. My initial need to "get a handle" on the school gave way to this profusion as I realized that any "map" of ESH would make it look stable and predictable. My own rhetorical performance of ESH bears the mark of those offered by girls, who constantly reminded me that innumerable readings of one locale could still be persuasive, and that those readings need not be tamed. But even still, this chapter bears the organizational mark of data analysis, coding, and editing. The traditional confines of academic research, as well as the stationary nature of text inevitably force a false unity on these competing understandings of the school.

Having set the stage for girls' identities, in chapters 4 and 5, I focus on how girls positioned themselves and were positioned within ESH's social world. In chapter 4, I focus specifically on how girls' used style to negotiate social scenes, groups, and cliques. Style was one of the key features that enabled girls to dis/identify with others, creating both cohesion and friction among them. And in chapter 5, I focus specifically on how girls used style to negotiate their "images," or their individual identities. Through the concept of "image," I also explore style as a manifestation of agency and girls' understandings of how they could alter the way they were seen by others—and by themselves.

Chapter 4

"If You Dress Like This, Then You're Like That": Positions and Recognitions

> If you dress a certain way, people judge you by the way you look, of course, right? Everybody does that. So if you dress like *this*, then you're like *that*. But I know I'm not like them. So I don't want to be perceived like them.
>
> Shen, "casual" style

While ESH presented its fair share of social challenges to girls, finding "your own people" was certainly possible if you knew how to look. Style was the most common way in which girls were able to see similarities and differences among themselves. As Shen declared in this chapter's epigraph, if you dressed like *this,* then you were assumed to be like *that.* In other words, girls actively worked to showcase aspects of their identities through the subject positions that style represented. When a girl noted that she "really liked" how another girl dressed, it often meant that she liked how that girl was performing her school identity, that she felt an affinity for that identity, and that she was also engaged in, or desired to be engaged in, a similar performance. Style was thus a serious consideration, as girls worked to make sure they wore the right *this* in order to be recognized as the right *that.* As Shen further explained, if you did not want to be "perceived" like certain girls, you had to ensure that you were not misrecognized because of your style.

This "complicated dynamic of recognition and misrecognition" (Gonick 2003, 11) facilitated the negotiation of school identities, as girls were positioned and positioned themselves in subject positions based on categories of race, ethnicity, class, sexuality, and lifestyle affiliations. These recognitions forged the identifications that produced identity, or "the detour through the other that defines the self" (Fuss 1995, 2). For Diana Fuss, identities are our "public personas—the most exposed part of our self's surface collisions with a world of other selves." Identification, however, is the initial process that enables us to enact the public performance of identity,[1] making identification the private realization of our deepest desire for belonging, acceptance, and connection to others (West 1995). Before an identification can take place, "some common origin or shared characteristics" (Hall 1996, 2) with others must be recognized. Conversely, through disidentification, we come to recognize who we are based on the subject positions we do not take up. Britzman (1997, 33) suggests that identification and disidentification are thus entwined in the struggle for self-recognition: "What is outside is also inside and this inside, call it difference, fashions both the dismissal of and the engagement in what can be imaged as the outside and the inside." The inside depends upon the outside for shape, definition, and validity, opening up the inside to perpetual renegotiations through its "founding repudiation" (Butler 1993, 3). This dependency means that identifications are continuously renegotiated and reconstituted. Identification is thus conceived as a "question of relation, of self to other, subject to object, inside to outside" (Fuss 1995, 3). For Fuss (1995, 2), identification is that which "inhabits, organizes, instantiates identity. It opens it up as a mark of self-difference, opening up a space for the self to relate to itself as a self, a self that is perpetually other." But as perpetually "other," identifications are never neatly experienced; they always bear the mark of ambivalence.[2]

At ESH, I met girls who desired to both occupy and not occupy subject positions at the same time. In these ambivalences, the possibility for keeping school identities in "play" emerged as girls worked toward maintaining the contradictory elements of their identities, without forcing those subject positions to conform to a tidy notion of the self. Gonick (2003, 13) suggests that ambivalence, as *the* marker of identification, offers "the possibility for expanding the horizon of these investments to embrace new forms of gendered subjectivity." Girls' ambivalent performances of girlhood, as they were expressed through style, created space in the school for a disruption of conventional girlhood, as well as additions and expansions to previously established notions of girlhood. This chapter is an exploration of the public subject positions that girls occupied through style and their relationship to the private and often uncomfortable process of dis/identification that enabled these subject positions to become intelligible to girls. In order to explore this struggle for inclusion and exclusion, inside and outside, I locate girls within various styles that existed at ESH. This range includes styles that girls named, as well as styles that I observed: "preppy," "alternative," "comfortable," "dressy," and "sporty." These broadly construed categories corresponded to fluid and shifting subject positions that girls occupied because they chose to, felt forced to, or more likely, a combination of both. But one other range of style deserves mention here as it infiltrated almost every discussion I had with girls: dressing "skanky." To say a girl was dressing like a skank was to say that she was showing "too much skin" and—more importantly—that she was showing it in the "wrong" way. Though only one girl articulated her own "skankiness" to me, it was a common insult leveled at other girls, operating as the perpetual outside to all insides.

The styles described in this chapter are by no means exhaustive. The range of styles that I have chosen to include (all research is, after all, a series of inclusions and exclusions) represents the most common modes of dress that girls brought up in conversation, as well as the most common modes of dress that I observed in the school. The categories of style that make up this range are ranges in and of themselves and are not meant to be taken as stable forms of embodied subjectivity. While girls positioned themselves within particular modes of dress that corresponded to particular subject positions within the school, they also expressed ambivalent feelings about their performances of these school identities. And finally, the styles described here are not isolated or self-contained. Each style existed in relation to the others, entangled in a web of mutual definition through the exclusionary matrices that operated throughout the school. Girls' ambivalent dis/identifications with particular styles and the shifting and fluid subject positions that these styles represented highlight the complexities and multiplicities of girls' school identities within the multifaceted social world of the school.

"DRESSING LIKE A GIRL": PREPPY, SEXY, AND POPULAR

The two acknowledged uniforms for girls—the Britney and JLo looks—were considered to be the most mainstream of all the styles at ESH. The trendiness and mass marketing of these styles meant that they were worn by the majority of popular girls, defined as those who knew the most people and thus wielded the most power in the school (Bettis and Adams 2003; Eder et al. 1995; Lesko 1988; Merten 1997; Milner 2004). As the popular girls at ESH were variously located across the school's multiple programs, as well as racial and ethnic groups, the uniforms had a great deal of variation. For example, many popular Asian girls, including "Nammers," "Hongers," and "Chinks," wore the velour JLo "suits" in soft, pastel colors, shopped at stores like *Aritzia*,[3] and wore labels, such as TNA, 555 Soul, and Kappa. Within this "hardcore" so-cial scene, popular Chinese girls who went to China with their families brought back other variations, such as tops with the words "Baby" and "Booty" written on them in Chinese lettering. And popular Hongers wore another modification, including loose fitting tops and baggy jeans that were still quite "revealing." However, these variations were hardly noticed by other racial and ethnic groups, and the JLo "suits" were seen as the main style for popular Asian girls, a style that was considered to be *the* most fashionable in the school. As Azmera, a girl in beauty school, explained, "if you're a girl, and you want like, a good reputation with the girls, you kind of just have to like, you have to dress good. You have to keep up a high standard." This standard meant "dressing like an Asian girl."

Popular white girls, often exemplified by the loud and quiet popular Frenchies, wore low-rise jeans and tight tops, bought name brands, such as Mavi, Miss Sixty, and Guess, and shopped at *Below the Belt, American Eagle Outfitter,* the *Gap,* and other stores that catered to a "chic" skater aesthetic.[4] Because both popular Asian and popular white girls wore each other's uniforms sometimes, the real difference between Asian pop-ular style and white popular style seemed to hinge on a display of skin, though the answer to which group displayed more skin shifted depend-ing on who I asked. Xiu's distinction between the hardcores and the Frenchies depended on the midriff and other subtle differences.

> The typical white group I see at school would have their pants all the way down to almost where you may see their underwear or lower. They like to wear the furry hooded jackets, tight light blue jeans and also, I'm guess-ing they are called skater shoes. That is fairly different from the Asian group because the Asian group does not like to show their belly that much. They also wear the hooded jackets with the fur, but their shoes are different. The Asians would go for the Nike Shocks instead of skater shoes.

Dren's knowledge of how popular girls dressed at ESH was similarly linked to the display of skin. As she put it, "pants are so low that you can practically see the hair line." But she felt that the popular Frenchies "don't do it [show skin] quite as much as like, some other popular girls. They *do* it, but not as much. I don't think they do it as much as like, I hate to say it, but *a lot* of the Asian girls are *really* into that." This dis-crepancy meant that there was a fair amount of finger-pointing over which popular group was more "inappropriately" dressed. White girls were viewed by some Asian girls as "trashy" for wearing their pants "all the way DOWN!" while Asian girls were viewed by some white girls as "hoochies" who wore their pants too "TIGHT!" These judgments were how girls articulated dis/identifications and served to create solidarity within each social scene, group, and clique.

However varied popular girls' style seemed to be, these outfits were collectively known as "preppy," a historically shifting category that is often synonymous with popularity (Merten 1997; Milner 2004) and refers to "going with the trends" and dressing like a "typical" girl, whatever that cultural construction happens to be at the time. Dressing preppy usually means wearing trendy or name brand clothing that marks students as popular within the symbolic economy of the school (Eder et al. 1995; Milner 2004). Maria, a popular Frenchie, articulated this distinguishing characteristic: "I try and be fashionable, and like, I know what fashion is, and stuff like that." The brands that Maria "knew" came from high profile stores that catered to a mainly middle-class clientele. Bettie (2003, 16) describes preps as "primarily white students who were the most middle-class of all students at the school." But at ESH, "preppy" was a more complex subject position that related to the school's postmodern milieu. While some white, middle-class students were preppy (though not all), including the popular Frenchies, the majority of the girls who dressed preppy were working-class Asian girls known as "hardcores." To be a hardcore was to swear "all the time," drink alcohol, go to raves and all-ages clubs, and associate with the gang-oriented Nammer boys.

Significantly, this notion of preppy represents a radical departure from how popular girls have been characterized in the past. Preppy style has never incorporated a "sexy" and "revealing" look until this particular historical moment. Instead, preppy has usually meant that a girl dressed modestly, conservatively, and in a manner that expressed femininity without sexuality (Bettie 2003; Eckert 1989; Lesko 1988). Preppy has also been synonymous with middle-class femininity and was incorporated under the overarching category "Jock." As I noted in chapter 3, at ESH, the binary opposition that Eckert (1989) identifies, namely "Jocks" and "Burnouts," did not exist in such a clear-cut or oppositional manner at ESH. In Eckert's analysis, the Jock/Burnout categories are "embodiments of the middle and working class, respectively; their two separate cultures are in many ways class cultures; and opposition and conflict between them define and exercise class relations and differences" (5). Yet, at ESH, a working-class girl had access to the preppy subject position. In order to afford preppy clothing, many hardcores worked part-time jobs at McDonald's, ice cream shops, Starbucks, and other local venues on Wellington Ave. Jade, a working-class hardcore, had a part-time job at McDonald's in order to afford a trendy bomber jacket with faux-fur on the hood, skin-tight Mavi jeans, Nike Shocks, and TNA suits in a variety of colors. Girls also received money from their working-class parents in exchange for doing well in school and for being a "good girl." As Maria proudly announced, "if I wanna go buy a pair of shoes—my mom will be like, 'Well, how much do you want?' And if I want a $100 my mom will just give me a $100."

Maria's access to her parents' money was, however, not the case for all working-class girls or working-poor girls, particularly those who came from single-parent families, like the girls in beauty school. The financial investment that the subject position "preppy" required thus limited access to those who were able to afford it, were able to work for it, or, as Xiu mentioned, were willing to steal for it. It was a subject position that indicated a girl had money to spend, though money that she perhaps struggled to get. Even still, as Azmera pointed out, the look was a "rich" one, and one that she felt looked "richer" on Asian girls than on white girls. She explained, "you could have a white girl and an Asian girl wearing the same exact thing, but it just looks different on them [Asian girls]. Everything just looks so elegant on them." This perception related to the "chopstick" thinness of most hardcores. The thinner a girl was at ESH, the preppier she could be. The relationship between preppiness and thinness had much to do with the fact that stores carrying preppy articles of apparel only stocked items that were in particular sizes. A plus-size girl could not walk into *Aritzia* and buy a plus-size suit. It simply was not available.

If the preppy style could be acquired and, of course, was *desired* by a girl, then it offered her entrée into a particular subject position within the school. It was not just that a girl was viewed as popular in this style (a girl could be popular without being preppy), but rather that she was viewed as being a particular "kind" of girl—a girl who looked "good" in the tight and revealing clothing that the preppy style necessitated, a girl who looked "sexy," and a girl who wielded power within the heterosexual matrix of the school (Hey 1997), where preppy girls set a "normative assumption" (Butler 1993) for femininity against which all other girls were measured and measured themselves. Preppy girls were seen to emulate the "cultural ideal" that was disseminated through movies, magazines, and music videos. Shitar wondered why "everyone says, 'Real people don't really look like that [like in the magazines]!' But you come to our school and you see that people really *do*. Not all of them, but the ones that are important do." Dressing preppy thus automatically linked girls to a particular body type, sexuality, and sexual power. Put another way, preppy girls were most likely to be pretty, thin, "hot," and involved in a heterosexual dating scene with popular boys.

Shitar explained how this "equation" worked: "Basically, what you look like is how you dress, like, there are—things, like, if your face looks like *this, this* is probably how you're going to dress." If you are blonde and pretty, she suggested, you were probably going to wear "Mavi jeans, low-cut, maybe show your stomach a little bit." If you were "a little large," you were going to wear a style that covered your body, "maybe sweat pants, nothing tight, you know, light colored." Diane, a plus-size girl, confirmed this correlation: "If you're skinny," she said, "you *can* be in style." For Diane, preppiness had everything to do with not only being able to afford the "right clothes," but also being able to *fit* into them. As far as she was concerned, preppiness, and all that the subject position entailed, was strictly off-limits to her as a plus-size girl, making other components of the subject position off-limits too, such as conventional sexiness and the kind of power that came with regulating femininity within the school.

To successfully occupy the subject position "preppy" was to be recognized through the mainstream, heterosexual matrix of "emphasized femininity" (Connell 1987, 183), or "the maintenance and practices that institutionalises men's dominance over women." The emphasized femininity of the preppy style demanded that a girl maintain a level of sexiness and thinness that was regulated by mass media, particularly pop stars, such as Britney Spears, Christina Aguilera, and Jennifer Lopez. While popular girls were often highly aware of the pressure to look "good," stay thin, and attract boyfriends, they also continuously alluded to the fact that they were caught up in something that was "really gross." Leah acknowledged that being preppy held her to a standard of femininity that was unrealistic. "It's like, it's almost been like, imprinted in your brain by like, everything! It's like a little tattoo that you can't get off!" Leah was referring to the pressure she felt to be thin and pretty based on the clothing that was currently available to girls, as well as representations of women in fashion magazines. She was able to articulate the "impossible" criteria to which she subscribed, while still acknowledging that she would "rather look like an anorexic person than an obese person."

As a quiet popular Frenchie, Leah's contradictory identification with the preppy subject position was made even more ambivalent by her fear of judgment from the loud popular Frenchies. She explained that if somebody walked by the loud populars, "they'll be like, 'Oh my god! That girl's trying to look like Avril Lavigne!' Or they'll just take someone and say, 'What are they *thinking!*'" Had it not been for the judgment of the loud popular Frenchies, Leah would have experimented with a different kind of style, something more "random," more "second-hand,"

and not quite so mainstream or explicitly feminine. But the thought of making a "full" transition away from conventional style worried her.

> I don't like being judged and I know there's a lot of people who, just kind of think that you are being like, a wannabe, or just trying to—you know, where there's a point where you think, "Well, I don't think I could pull this off," which is kind of stupid, because you can, in theory, pull off anything. I've kind of begun to realize it this year, but I don't want people to, you know, make fun of me or relentlessly tease me or get angry at me because I'm wearing something that they don't think I should be, so—.

Leah's fears about being judged by others were felt by girls across the school. Girls worried about the exclusions that could be exercised upon them and surrounded themselves with girls who would not judge them, because they, too, wore similar styles.

Maria offered an example of judgment by articulating her own constituted "inside" against the constituting "outside" that "skanky" girls provided. As a thin, pretty, working-class, Italian girl in the French Immersion program, Maria wore JLo suits and *Aritzia* fashions on most days, acquiring the nickname "Barbie" for her "girly" style. As she often said of herself, "I love being girly!" When I asked Maria to define "girly," she explained it as, "Um, pink, pretty, um jumpy, excited, and like, ditzy at times." Maria did not have a problem with being recognized as girly and thought of it as a "natural" way for her to be that made sense given how she saw herself. "Just comes, kind of. Like, I'll just—people will just be like, 'you're the girliest girl I know!' And I'm like, 'well, I can't really help it, kinda thing.'" Maria's normalization of her own emphasized femininity created a familiarity between her and other preppy girls. It also gave her the power to judge girls whom she deemed to be less "appropriately" dressed.

> I think if you see someone dressed like Alice [her working-class, Vietnamese friend] and me, no one says anything, cause I think everybody dresses like that. So, it's *normal*. But if someone came to school, with a short skirt and high platforms, flip-flops, and panels in their hair, then, I think that everybody's going to be like, "Oh my god!" And there's going to be like, a lot of gossip.

Maria's validation of her own style as "normal" exemplifies the power that preppy girls had to regulate femininity in the school. While not all girls felt compelled to adhere to this proper citation of femininity, the "normal" style set by Maria and others in this subject position was seen as the benchmark for an "appropriate" performance of girlhood, making all other dressed bodies "abnormal."

No matter what other style a girl wore, the preppy style infiltrated it, shaped it, and infused it with ambivalence. Ironically, the "skanky" subject position also held a similar power. It shaped and infiltrated all other styles, too, forming perpetual disidentifications that necessitated guarding against this damaging label. Preppy girls had the power to apply this label *to* other girls. I constantly heard hardcores and popular Frenchies refer to girls as "skanks." It was, in fact, the most common example of judgment in the school. But, interestingly, preppy and skanky styles shared a threshold of "sexiness." As I have noted, both preppy white girls and preppy Asian girls often considered each other to be "skanky." Dressing preppy and dressing "skanky" were thus closely related. As Gianna, who sported a comfortable JLo look, said, "Oh! There's a very close line between looking good and looking like a ho." But this line was impossible to determine and depended upon how others looked at a girl and *not* how a girl thought she looked. In order to be skanky, a girl had to be "*too* sexy" in her style; in order to be preppy, a girl had to be sexy in her style, but not "*too* sexy." These lines were arbitrarily drawn and constantly shifting. While "skank" was an insult that preppy girls leveled at those who were seen to be "different" from them, it was also an insult that they leveled at each other as a way of regulating power in different social scenes, groups, and cliques across the school.

For Chrissie, the ambivalence she felt around her preppy status was infused by what she expressed as a misrecognition of her style. People thought Chrissie was skanky, even though she wore what *she* thought were "normal" clothes. "I never thought, 'Oh, if you dress *this* way, people will think *this* or *that* or *that*.' Like, I thought, it doesn't matter how you dress. And then, I kind of learned, slowly, 'yeah, it does!'" Initially, Chrissie described her style as preppy, which she quickly amended to "skanky almost or hoochie maybe." But she told me that she no longer "had that style anymore" and that she had begun gravitating "more towards baggy clothes." When I asked her why she decided to switch her style, she recounted how others were "getting this impression of me, that I was like, a preppy like, hoochie kind of girl, who went out looking for guys, dressed in tight clothing, or whatever." In her "tight" miniskirts and "skimpy tops," she noticed that girls were "just very blatant; they're like, 'Tramp!'" Chrissie, unaware that her style was sending out "skanky" messages, suddenly noticed "enough little comments that I was like, 'Oh my god. People think I'm someone I'm not at all.' And it really started to bug me." As a thin, pretty, blonde who drew attention from boys everywhere she went, Chrissie was subjected to the judgment of other preppy girls, who called her "anorexic," "bleached blonde," and "that damn Britney Spears." As Chrissie explained, some of her friends had even started excluding her from social events, fearing that she would monopolize the attention of boys. Based on these experiences of exclusion, she noted that preppy girls—her friends—"are quite judgmental."

When I met Chrissie, she had just cut her hair short and dyed it a light brown. She had also recently gained some weight. At school, I occasionally saw her wearing hoodies with the hood pulled up over her head, cargo pants, sneakers, and slightly baggy sweaters—a skater style that she had begun to cultivate, along with her more "regular," preppy look. I asked her if her new look had anything to do with the way other girls were judging her.

> Yeah, it all kind of adds up. Like, I don't want to be some Britney Spears skank, blah, blah, blah. I don't want to be known as wanting to go out and get attention from guys. I don't like any of that stuff that apparently like, I guess like, my overall image presented to people.

Though Chrissie had developed a skater style to counteract her "skanky" image, toward the end of the year, I saw her wearing outfits from her "preppy" wardrobe more and more. As we sat next to each other in English class one day, I could hear Christina Aguilera's latest hit, *Fighter*, blasting from her disc player. Chrissie admired Aguilera's ability to be sexual, powerful, and to fight against what Chrissie had called "suppressed female sexuality." She also identified with Aguilera's recent image change from "just another pop star doing whatever" to announcing,

> "Yeah, I'm not a virgin, blah, blah, blah," like, everyone can deal with that, and sorta just, you know, even if it is slutty, she's not denying it. Know what I mean? And like, I always hear guys saying it about, like, um, using girls for sex and stuff like that. And what she's doing is just coming right back and saying, "Yeah, *we* can do that too!"

As I listened to Aguilera's commanding voice through Chrissie's headphones, I noted that she was wearing a white miniskirt with a deep slit, a tight, pink T-shirt, and strappy sandals with three-inch platform cork heels. Her hair had become a little more blonde and she had recently acquired a nose piercing similar to the one Aguilera sported on the cover of *Rolling Stone* magazine that month. It was clear that she had not given up her "old" style. Chrissie enjoyed the power of her sexuality, a power that she equated with expressing herself and challenging the double standard of male sexuality. But at the same time, it was a power that generated envy from others and, sometimes, resentment from her friends. Yet Chrissie's incorporation of two juxtaposing styles—the tightness of the preppy look and the bagginess of a more casual, skater look—represented her own struggle for embodied subjectivity within the social world of ESH, and her desire to be ambivalently located within and between these subject positions.

Chrissie also received a lot of criticism from "alternative" girls in the school, like Dren. Dren argued with Chrissie about the validity of Aguilera's so-called "power," as well as the validity of Chrissie's preppy look. Like other girls who called themselves alternative, Dren's style stood in (ambivalent) opposition to preppy style, creating exclusionary matrices that disidentifications with subject positions necessitated. Of course, alternative style begs the question, "alternative to what?" At ESH, the "what" was clearly defined as those girls who dressed "like everyone else, so you really need to hunt them out. You need to sniff them out and destroy them." When I asked Mina whom she was referring to, she replied in a quick and assertive tone, "preppy girls of course."

"I'm Nothing Like You": Alternative, Goth, and Punk (Ass)

Jamie called herself a Wicca and was never without the cat's-eye glasses that sat slightly akimbo on her face. She wore loose fitting jeans, tops, and nonskater (read: untrendy) runners. Jamie sported two visible piercings in places that were highly uncommon for girls at ESH: a stud between the lip and the chin (the labret), and a barbell through her tongue. Her other piercings were hidden within her mane of long, brown hair. She had two in one ear and three in the other. "Yeah, and belly button! But the belly button got infected so I had to take it out. But I still have the hole so it still counts." Because there were so few visibly pierced students at ESH, those who sported facial jewelry were easily recognizable to each other. Jamie's boyfriend, Dante, had pink hair and nipple rings that he had inserted himself. "If people want to see them," she told me, "he'll just lift up his shirt in the middle of nowhere." They met when Dante transferred into Jamie's math class. Dante saw Jamie and knew right away that he was going to talk to her. "He goes, 'she's my favourite' and then he walked in and sat behind me. So that was the first thing he saw—my piercings."

While Jamie's piercings enabled Dante and others to identify with her, they also generated judgment from preppy girls. In order to make her uncomfortable, some of the loud popular Frenchies sometimes asked Jamie "embarrassing questions," such as "does your tongue get stuck in

Dante's nipple ring?" This question implied that Jamie was involved in an "aberrant" sexuality that revolved around an aggressive desire to use her tongue piercing to inflict pleasure and pain on Dante. As well, girls with tongue piercings were seen to be "into" oral sex. But her tongue piercing was "not for *that*," she insisted. "I don't want to be seen that way. I want to be seen as *different*."

Being recognized as "different"—no matter what the social penalty—was the key criterion for girls who occupied the "alternative" subject position at ESH. While the preppy subject position crystallized around mainstream style and the power that came from enacting an emphasized femininity, the alternative subject position was based on a girl's refusal to do just that. As a result, no two girls who thought of themselves as alternative dressed in the same way. At ESH, being alternative meant being "one of a kind" and carefully cultivating a look that could not be misrecognized as preppy. Alternative girls did not wear labels, did not show skin, and did not conform to the Britney and JLo uniforms. Their diverse styles served as a visible foil to the range of preppy styles, including normalized body types and the heterosexual imperative that dominated the school. While alternative girls had alternative boyfriends, many of them also expressed interest in bisexuality, or as Mina put it, "I'm lesbianesque with a hint of hetero." Alternative girls also tended to be boldly sexual without being conventionally sexy. Mina, Jamie, Dren, and Ratch had all slept with their boyfriends during the year, while they felt that preppy girls were "prudish," or that they "did it" but refused to "talk about it." Not only did alternative girls seem to talk about sex, but also they often used their sexual knowledge to make others uncomfortable. Mina brought a Barbie and Ken doll in to class one day and arranged them in "dirty" poses on her desk. She had dressed them as "gay Ken" and "Butch Barbie," drawing constant attention to their gender-bending sexual exploits. I also heard Dren and Ratch bring up masturbation in class numerous times in order to get a "rise" out of the preppy girls, whom they felt were "really close-minded" when it came to talking about sex and female pleasure.

Alternative girls called themselves "different," "weird," "not normal," "unique," "punk," and "goth." Jamie's best friend Mina saw herself as alternative. Her style was purposefully unkempt and "grungy." She often combined a loose-fitting, hot pink tank top with her omnipresent black trench coat and man-sized jeans that dragged under her feet. Jamie described Mina's look as "different," because she had her "own style going on. Sometimes she wears pink and black. I mean, with somebody who wears black, you don't usually see them in pink." Mina's alternativeness was also reflected in her interest in Japanese anime, Wiccanism, and her aggressive (some described it as "psychotic") way of dealing with people, particularly boys. She was loud, but not in "that preppy way." As Jamie explained it, "Mina can be at the other end of the hall and scream her lungs out at you and not care what people think about her." She also "play-fought" with the boys in her social group, biting them, scratching them, and lunging at them for the slightest provocation. Making others uncomfortable was Mina's speciality.

Mina explained that she and Jamie were friends because neither of them was "flaky," like the preppy girls, who spent

> hours talking about which clothes go with which makeup. It's just really annoying to hear them talk because it's like, where's this going to get you in life? What are you thinking about doing? I'd rather have philosophical debates. I do that all the time with Cedric [her boyfriend] and Jamie. That's why I fit in with them.

Mina and Jamie's identification with each other rested on their mutual distaste for mainstream girlhood, as well as their mutual desire to perform an alternative one. According to Mina, an alternative girl was "someone who wouldn't do just *anything* to fit in, someone who would rather be their own person than go with the mainstream." Mina proved her commitment to this ethos by bragging, "people have tried to get me to be popular. They've invited me, right? And I don't WANT to!" She judged the popular Asian girls for being "giggly," "fake," wearing "the same kind of makeup that looks like it's been painted on," and dressing like "hoochie mamas," who "wear heels up to here! Their thighs are like *this* wide, they're wearing really tight jeans and really short skirts." Similarly, she judged the popular white girls for hiding "behind masks" and pretending to like things just because they were trendy, such as "being in love with Justin Timberlake."

I asked Mina why she thought those girls were popular. "They tend to be popular," she replied, "because they're exactly like everyone else, and if they're exactly like everyone else, then you can't exactly hate them, because they're exactly like you, and if you hated them, you'd be hating yourself!" Though Mina's assessment of the preppy subject position was grounded in her critique of mainstream society, it was ambivalently experienced. While Mina and Jamie hung out in the isolated privacy of the drama studio during lunch, keeping themselves on the fringe of ESH's social world, Mina battled her own uncomfortable desire to be recognized by preppy girls, who saw her as a "freak." "I feel like I'm on parade each time I walk across the cafeteria, on my way to my bench for gym," she explained. "I'll just feel that everybody's eyes are on me." When I asked

Mina what she thought the preppy girls were looking for, she suggested they were "looking for perfection."

> I'll feel that maybe my chest isn't big enough. Maybe my arms and thighs are too big. Maybe my stomach just hangs out just a little too much. Maybe my butt's too big. Maybe my hair's greasy. You know? And if any of that is true, even in the slightest, [preppy] girls tend to exaggerate it. They take it into their minds and they conform it to see this huge, monstrous imperfection.

Mina's performance of an alternative girlhood was thus infused with a desire to critique preppy girls and to be accepted by them as well, but on her own terms. While she was not willing to compromise her look or attitude, she continued to hope that they would find her pretty, witty, and intelligent. Jamie told me that she, too, did not want to be totally unaccepted by preppy girls. Her alternative style and attitude were "different," but also "socially acceptable," or, as she explained, "to a lot of people 'different' means you're NOT socially accepted. I'm different in a way that people see me as being different from other people, but they *like* the differences."

The styles that existed within the alternative subject position ranged from Mina and Jamie's more subtle differentiations to Dren and Ratch's subcultural affiliations with goth and punk respectively. Somewhere in the middle were Abby and Shitar, two girls in the French Immersion program who, though they were not friends, shared a healthy dislike of preppy girls and an affinity for putting together unique outfits that stood out in the school. Abby was not a firm part of any social group at ESH, though she had some friends with whom she "roamed the school" at lunchtime and talked about the things that interested her: Orson Welles, the *Lord of the Rings* trilogy, and philosophy from Plato to Nietzsche. Abby defined her style as "sort of a bad-ass-schoolgirl-business-woman look." She combined high-heeled Mary Janes with pleated skirts and her trademark striped kneesocks. As a working-class girl with no options for purchasing expensive fashions, Abby was a bargain hunter who looked for second-hand clothing at *Value Village* and in lost-and-found boxes around the school. "I just wear what I want to wear," she told me. "You know, like, if I think it looks neat, then I won't really care about what everybody thinks about it or whatnot."

Abby had only recently chosen to occupy the alternative subject position at ESH. Prior to discovering the striped socks that changed her style, she was a "wannabe popular girl." "I used to want to be [popular] so badly, you know?" In an effort to become popular, she felt that she had to dress just like preppy girls, in tight jeans and tight tops. But she called striped socks her "salvation." When she saw the socks in a store, "it was like in the movies, where light comes down and it's like, '*Ahhhh!!!*' like, '*I love you!*'" Abby identified with the subject position that she felt the socks could offer her, a unique standing in the school that would signify her shift away from "wannabe popular" to a "weird" girl with her own style. She then stopped trying to fit in with the popular girls, or, as she put it, "I just started not to care." Abby developed her own critique of preppy style and the "cattle farm mentality" that she felt was bred in the school.

> It's the most disgusting thing when you have two, like three girls walking down the street and they'll all be wearing tracksuits like that, and every tracksuit is like, the same color almost. Yeah, like baby blue or baby pink or, I don't know, brown or green or. God! I don't know. It makes me sick! Like, three girls are walking down the street and they all have a shirt that's baby blue and has "Bootylicious" on it! That's disgusting!

Though Abby's dedication to occupying the alternative subject position was strong, she also expressed ambivalent feelings regarding her inability to wear the "right" clothing based on her class location and her body type. She told me that she would never wear Mavi jeans, "which is what *everyone* else is wearing." When I commented on the cost of the jeans, Abby responded, "Yeah. Like, I don't have the money to spend $70 on a pair of pants that don't *flatter* me anyways." Further emphasizing her point, she added, "like, they *really* don't work on me." As a slightly larger girl than the "chopstick" thin preppies, Abby was well aware that in order to be preppy, a girl had to both be able to afford the "right" clothes *and* look good in them. As Abby felt that she could not do either, I asked her to what extent wearing the "right" clothing mattered at ESH. "Well, some people really care! Like, people do judge you by it like, on the streets." Abby told me about the time when she and another alternative friend were standing at a bus stop with one of the loud popular Frenchies.

> We were with Shelley and she was wearing her little tight jeans, and her like, one of those hooded sweater things that are "in" right now or whatever. And these girls came up and started trying to pick a fight with us, and they like, completely left Shelley alone, like she wasn't even with us, you know, it was just us two.

Abby felt that the girls had picked on them because, "well, I had my striped socks on and [the other alternative girl] was wearing some darker makeup." When the girls started to push Abby and her friend around, she took out the pepper spray that she always carried with her and they "backed off."

For Abby, the recognition she sought as an alternative girl was linked to exclusions like these that made her an outsider in the school. But she did not seem to mind. Her style took work and planning, and to her, it was worth it. Though she was not popular and did not have a large circle of alternative friends like Mina and Jamie did, she managed to find her own people here and there. She sometimes hung out with Leah, who was working to shift her preppy subject position toward something a little more alternative, though she was not yet prepared to go to the same lengths that Abby had in order to be recognized as "different" in the school. Unlike Leah, however, Shitar felt that she had nothing to lose by being alternative. As she aptly put it, "I don't have anything else. Like, I don't have a big talent or anything. I don't have a sport I love, or, I'm not really good at school. I don't have millions of friends. I have my style. I can make clothes."

When I asked Shitar what drew her to the DIY aesthetic that she called the "punk ass" look, she emphatically responded, "It's not zombie-like. It's not, 'Everyone's the same, let's all be a nice big family of zombies!'" I asked her who the zombies were. "The zombies," she heatedly replied, "are Gap shoppers. Anyone who—anyone who wears the tight jeans, or the like, tight halter top with, makeup and—that's a zombie!" Shitar's look, ironically, was zombie-esque in the sense that she culti-vated a slightly goth, slightly punk, mainly DIY aesthetic. She wore mix-and-match styles that she seemed to make up out of anything and everything, including jeans that had been cut short and shredded at the bottom, cow patches, studded belts, sailor shirts, striped socks, stapled black converse sneakers (a real distinction from the pastel and sparkly runners of hardcores and loud popular Frenchies), dyed-black hair, homemade jewelry, spiked wrist bands, men's ties, and safety pins that held it all together.

At the beginning of high school, Shitar worked to become friends with the loud popular Frenchies in all of her classes, the ones who, in Shitar's words, "think they're Britney Spears." She tried to dress and act "like them," even losing twenty pounds through the anorexic guidelines laid out on pro-ana Web sites.[5] But no amount of effort generated recog-nition from the loud populars and she remained an outcast. She realized that she could not afford the Mavi jeans and Guess shirts that were a necessary part of being popular within the French Immersion program. Her working-class background, as well as her Lebanese culture and strict

Christian upbringing distanced her from the middle-class, white preppy girls in her social scene. As a result of these differences, she articulated a sentiment most alternative girls expressed: "No one gets me," she sighed. "None of them are really like me. They have their lives. I have mine."

This disidentification propelled Shitar in a new direction. She developed a critique of the preppy subject position as "conformist" and decided that she liked her "spot" as an outcast in the school. If the "blondes" would not accept her, she reasoned, then she would also reject them. She started to notice, "Other people, like, who are like me, who stick out." Her own style started to develop through this identification with alternative girls, both in and out of the school. "When I go into certain stores, or downtown," she remarked, "there's all sorts of weirdos." As a "weirdo," Shitar took comfort in knowing that there were others like her out there, "strange people" whom she felt were kindred spirits. Though, she admitted to feeling an occasional pang of regret when she saw the "normal" girls,

> with like, the straight blonde hair and the tight pants. They all have boyfriends. They all have great lives. They like, party all the time and, and then I'm not like that, and I think, if I just dress like them, maybe I would have that too. And then I think, I've tried that and it didn't work. So, I don't know. *They're trying to draw me in!*

Shitar's obvious ambivalence made her laugh. I asked her if she ever considered giving up her alternative style in order to be preppy. "I consider it. But then I think to myself that it would be wrong." Shitar felt that the alternative subject position was much more a reflection of who she really was.

Adding insult to injury, Shitar's alternative style was not admired by some of the other alternative girls in the school. To Ratch and Dren, Shitar was a "class-A twinkie." A twinkie wanted to "get into the cool group," but did not have the authenticity to "pull it off." The cool group, in this case, was not preppy. It was, in fact, a group consisting of Dren and Ratch, best friends who recognized in each other a desire to occupy the alternative subject position as confidently and as powerfully as possible. Twinkies were those who did not have an "authentic" commitment to being different. They were merely posers who only seemed alternative on the surface, but did not embody the "true" spirit of punk, metal, goth, or other subcultural affiliations. "It's just like, these kids," Dren explained, "they shop at *Cheap Thrills*, or BUY the bondage pants, and they'll BUY the bondage skirt, when you're supposed to MAKE the bondage things." Ratch added, "you've got to put the time into it. You've gotta develop your look. It takes a while. You can't just go to the store and buy your fuckin' look, right?" When I questioned why Shitar was a twinkie if she, too, made her own clothing and had a DIY aesthetic, Dren and Ratch looked at each other and winced. "She just tries way too hard," Dren said. They felt that Shitar's look came from *Seventeen* magazine, where they tell you how to "make cool clothes by yourself!" Dren mockingly read an imaginary headline: "Oooh, cut your pants like *this!*"

As harsh as these sentiments were, they represented Dren and Ratch's exclusionary matrix, the outside to their self-proclaimed alternative inside. Though everyone called Dren a goth, she actually thought of herself as both goth and metal, alternating between these two very different styles. Her goth look consisted of tall, black leather boots, flowing, black gossamer shirts, dyed black hair, and spiked bracelets and dog collars. She noted that goth girls were "very feminine, but they're like, dark." Metal, however, was quite different, though no one seemed to notice the difference. Her metal look included tall, black leather lace-up combat boots, jeans or cargo pants, chains, a black army jacket, an Iron Maiden T-shirt, and the facial piercings she sported year-round—a much tougher, more masculine aesthetic than her goth look.[6] And Ratch was a tall and strong looking punk who wore army pants decorated with graffiti, safety-pinned patches, punk band T-shirts, ripped hoodies, dog collars, and had long and messy hair. Unlike her estimation of Shitar, Ratch felt that she authentically embodied the punk spirit of DIY.

> If you're confident with who you are, you can pull off anything! Like, people are like, "Wow, a lot of people who do that, a lot of people can't pull it off!" But it's just being comfortable. It's being confident in what you wear. Cause if you dress like us, you have to have a lot of confidence, or people are going to be like, "What the fuck are you doing?"

The unique feature that allowed Dren and Ratch to judge others so harshly was the fact that they had come to ESH with an already established alternative look. They were not outsiders who were denied access to the preppy subject position; they were not even unpopular. In fact, Dren and Ratch understood that they had been able to remain popular "on their own terms," simply by "doing exactly what they wanted," a feat that other girls often admired. While most alternative girls were drawn to the alternative subject position because preppy girls rejected them,

Dren and Ratch never experienced such exclusions. The summer before high school began, Dren simply decided to start "totally fresh."

I mean, that summer I went out and bought a whole bunch of different clothes. I was like, so determined that I wasn't going to go into high school looking like all my friends, and showing up on the first day and getting into fights with your friends because we'd all bought the same shirt.

By grade ten, Dren had cultivated her look, including an eyebrow ring, a lip ring, and a dedication to goth and metal music that gave her an intense credibility in the school. She read goth and metal music magazines, went to all-ages goth dances, and was never without her disc player in class, which blasted industrial rock, as well as death and black metal—her large and powerful headphones perpetually sticking out of her long black hair.

Ratch's style also grew out of her taste in music. "I found a music that I liked, and then I went with the dress." Like Dren, she experienced notoriety at ESH for her look—attention that was sometimes based on the fear she generated as an intimidating figure. "When I walk down the hallways, I say 'hi' like, to everyone. But I know there's people who cringe when they look at me." Those people considered Ratch's style to be "dirty," "disgusting," and "street kid." As she put it, "this girl in my Socials class looked at me and looked at her friend, and was like, '*Gawwd, can't you like, get a job and go shopping?*'" The negative judgment that Ratch experienced and Dren did not related to their differing class locations and choices of subcultural style. As a middle-class girl, Dren was able to afford the expensive skirts, dog collars, and leather boots that the goth and metal looks necessitated. As well, her goth style was somewhat sexy; it did not seem to challenge the emphasized femininity of preppy style as much as Ratch's punk style did. For all her goth and metal accoutrement, Dren was still a thin and pretty girl. Ratch, however, was tall, strong, and masculine looking. Ratch was a working-class girl who lived in an apartment with her single mom and brother. Her inability to buy goth style was not an issue, however, as she preferred her chosen look and the DIY aesthetic that gave her authenticity in the school. As Ratch pointed out, punks were not supposed to spend money; it defeated the "whole point!"

At ESH, there were only a handful of girls who identified with subcultural styles, and only two who embodied these subject positions intensely enough to be recognized school-wide. This solidarity enabled

Dren and Ratch to identify with each other and share in their critique of preppy girls. As Dren explained,

> when you're going to school every day and you're getting up at eight, or seven or whatever in the morning and you're piling on half a bag of makeup and wearing skimpy little tops and it's like, autumn or winter, and you're showing up at school and like, pouting your face and you're going into the bathroom every half an hour to like, reapply your lipstick, just because you're like, in English class with a couple of dorks that are your age anyway, like, it's pretty stupid.

But as critical as Dren was of preppy girls and their desire to be attractive to boys, Ratch commented that she thought Dren was not so different from them. Ratch explained that Dren was "rather boy crazy, which kind of makes her like a lot of the girls I see around. I'm not saying that I'm *not* boy-crazy or whatever, but I'm not like that all the time and I don't have to constantly talk about guys all the time, which she seems to take joy in doing." Dren's heavily "made up" look caused others to speculate that she was similar to preppy girls, who also put effort into looking good and applying makeup. As well, Abby and Jamie's admiration for Dren's "outspoken" style diminished when they learned that Dren was "mean" and "judgmental," qualities that they associated with being preppy.

Even within a subject position predicated on "difference," girls judged others harshly for not being different enough or for not being different like they were different. There was an understanding among alternative girls that at least they were *not* preppy; that their styles said, "I am nothing like you!" Dren summed up this alternative ethos:

> Just to do your own thing and not really conform to whatever everyone else is doing. And not to like, shop at *Below the Belt* and *Stitches,* you know. That you, like, look different but not only looking different. Just that you think differently. That you're not totally like, obsessed with school, or not necessarily even that, but just that you don't necessarily do what everyone else does; that you kind of dance to the tune of your own beat. Yeah.

But doing "your own thing" was difficult. As different as alternative girls were from preppy girls (and from each other), they, too, formed tightly knit groups that operated on elements of exclusion. Mina and Jamie's social group had become quite "elitist." Initially, the group was drawn to the drama studio at lunchtime because they "were rejected by other groups." As a result, everyone who came to the studio had their own "thing," a thing that made them unique and different from everyone else. But eventually, everyone came to like the same comics, the same movies, and the same style. Mina jokingly recounted how hats had become the newest "thing" that all of her friends wore. "They're all supposedly alternative," Mina complained, "but alternative's now popular, so it's no fun anymore!" According to Mina, all the "things" that made everyone in the drama studio different had now become conformist. As a result, she admitted that her alternative group might have become just like all the other mainstream groups in the school who had rejected them in the first place.

"Show Class, Not Trash": Comfortable, Appropriate, and Dressy

For the girls who could not afford to occupy the preppy subject position, but also did not care to engage in the mainstream critique of alternative style, there was another option, one that was perhaps the most common subject position in the school. It included a range of styles that took its cue from the Britney and JLo looks, but did not incorporate labels or expensive fashions. As well, the girls who wore this style deemed it to be much "classier" than preppy style. Though it was made up of "no-name," knock-off fashions, it did not seem to reveal as much skin. Because of this distinction, the girls who shopped at "reasonably priced" stores, such as *Le Chateau, Stitches, Suzy Shier, Aldo, Transit,* and *Dynamite* saw their look as "comfortable," as opposed to the revealing, tight, and expensive clothing worn by the preppy girls. Keisha was sure that girls who showed their stomachs, thongs, and cleavage could not possibly be "comfortable." "I wouldn't be comfortable," she confidently asserted, "so I'm pretty sure they can't be comfortable. I just honestly think they do it for attention." According to Keisha, girls who wore the Britney look were sending out the wrong messages to boys, making them think "those" girls were "easy," "slutty," and only interested in "popularity." Girls automatically noticed when other girls dressed "inappropriately," she continued, "and it'd be like, "'sssspssssp' and talk about it! But *guys,* guys will notice it when they see it. And be like, 'Oooooh, look at her! Wow! She's showing a lot.'" As far as Keisha was concerned, "if you come with a shirt showing your stomach and your cleavage, guys are obviously going to think, 'Oh, she's easy! I can *get* some of *that.*'"

Keisha's adamant dislike for the Britney look was a part of the "comfortable" subject position and made her recognizable to other girls from a working-class and working-poor background. This identification incorporated girls who wanted to look stylish and fashionable, but did not have the money to buy the clothes necessary to occupy the preppy subject position, or simply did not want to. The girls in the comfortable subject position wore jeans, T-shirts, hoodies, sweat suits, and runners, but they were noticeably different from the not-altogether dissimilar styles of preppy girls. They were less trendy and less expensive. Comfortable girls sometimes showed their midriffs and their thongs, but most did not. Like Keisha, Shen saw the Britney and JLo looks as a desperate grab for "attention from guys, and like, maybe, you know praise from their girlfriends, right?" She mockingly imitated what preppy girls might say to each other: "'Oh, that's such a *nice* top! Oh, you look so *sexy!*' or whatever." Shen, a quiet, Christian, Chinese girl from an immigrant, working-class family, did not "approve" of the Britney look at all. "It definitely makes them [preppy girls] seem less classy, you know?" Similarly, Isabel made a plea to girls who wore low-rise jeans and midriff-revealing tops: "please don't go wearing pants that are just like, barely covering you and when you sit down and you've got to sit down holding your pants up, because if not, people are going to see, you know, your crotch or, your ass crack or something, you know!"

The most identifiable feature of the comfortable subject position was a critique of the labels that preppy girls wore. The mere mention of the JLo suits prompted Keisha to articulate her own class location as a girl whose paycheck went straight back home to help her family.

> I hate the labels that have to like, cost so much. And it's sad because, you know, not everyone can afford it. And so what if someone really wants a pair of like, Mavis or whatever the case may be, and they can't because it's too expensive for them, and then they get judged on it at school. And I don't think that's fair. That's why I think all high schools should be in uniforms. That way no one can get made fun of cause they're all the same.

Keisha's hatred of labels was shared by her beauty school friends, many of whom worked in part-time jobs, but needed the money to help out at home, pay for their beauty school tuition, or pay for their car insurance and gas. Jo, a working-poor girl who lived in the "Projects," also explained that girls were judged by how they dressed at ESH. This judgment was enacted by the "people who have more than others" toward those who wore styles that were not "name brand, or in style." Jo, who was troubled by the lack of solidarity that different styles sometimes created added, "We're girls! We should get along. Girls stick together."

Gianna took a different approach. Her theory was that even if she had the money to buy expensive labels, even if she were a "millionaire," she would never purchase the clothing that did not make her "feel like who I am, you know?" For Gianna, dressing like "who she was" meant being recognized as someone with a working-class, east side background. But, ironically, her adoption of no-name brand styles that resembled the Britney and JLo looks caused others to misrecognize her as a prep. A friend at another school accused her of looking preppy when she wore a particular pair of jeans. Her friend wore "spiked bracelets and like, the wallet with like, the chain and like that kind of style."[7] Gianna could not understand how she had been misrecognized as preppy and figured it had to do with her desire to "to keep up with the fashions." But to Gianna, being stylish was not a preppy characteristic, since her clothes were all no-name brands. Thinking of herself as a prep meant that she had to try to identify with the "rich" girls she knew at ESH and through her work at a spa. The identification proved to be impossible for her, "cause I see the west side kids that come to my work and I'm just like, 'No!'"

The comfortable subject position generated a sense of pride and honor among the beauty girls who felt that they were well dressed, and not at all as "inappropriately" dressed as preppy girls. When I asked Keisha what she liked about her style, she explained that it was the confidence that this pride gave her, knowing that no matter what she wore, she was "still comfortable." As a result, she did not care if "people are going to be like, 'Why's she wearing that?' Or talk bad about me. I have confidence, like I don't care! Like, I block them out. I'm in my bubble!" Keisha's self-assurance was shared by the beauty girls, who mainly dressed in styles that were similar to, but recognizably different from, the name brands that comprised the uniforms for girls at ESH. As a way of distancing themselves from a subject position that they could neither afford nor identify with, they spoke of themselves as "casual" dressers who liked to "look good," "tasteful," and "appropriate" for their age. Keisha very clearly explained that "the way I dress and present myself, I think it's the right way to present, for a 16-year-old, being in high school." Keisha took a moment to think about her style and then added, "I like to present myself a certain way. See, that's the difference, like, someone can dress very like, provocative, and look bad. And someone can dress in sweat pants and look prettier."

Shen wore fitted tank tops, but always made sure to put something over them, like a jacket. She noticed the way boys and men looked at girls' "asses" in their tight jeans and "suits" and did not want that to happen to her on the bus. But when Shen had the opportunity to buy a *Buffalo* name brand suit at a discounted price, she leapt at the opportunity. Articulating ambivalence with the comfortable subject position and her class location, she explained that she "really hated suits, but I actually got one a few weeks ago, and I was like, pretty mad at myself, cause I felt like such a hypocrite." Shen desired to look "good," but then admitted to hating the "*cool* people" with whom she would now be associated. But, the thought of this possible misrecognition did not deter her from wearing the outfit.

While the beauty girls took pride in their comfortable styles, they also worried about being misrecognized as girls "without class" (Bettie 2003), given their vocational positioning within the school. Keisha worked hard to maintain a "classy" style, "cause that's how people see you. You know, that's how people judge you, by the way you present yourself. And if you present yourself the wrong way, then people are going to be like, automatically, 'she's not a good girl. She's a bad girl,' you know?" To Keisha, being a "bad girl" meant doing "wrong things, like especially with guys. Like, trying to get guys' attention. Maybe, like, fool around." But Keisha maintained that "if you present yourself a good way, then [people] look at you, 'Oh that's a strong young woman, that goes to high school, that knows what's right from wrong!'" It was imperative to Keisha that she be perceived as the "right" kind of girl who was not misrecognized for being "skanky." For Keisha, this meant making sure that her friends also did not come to school dressed "inappropriately." And if they did, she would tell them, "cause they know like, that's *not* the look to go." The family mentality of the beauty girls meant that they all looked out for each other, cared for each other's reputations, and made sure that they watched each other's backs (and thongs). If a beauty girl came to school in an outfit that was deemed to be "skanky," "we'd all start talking about it, and be like, 'Why is she wearing that?' And give looks and whisper."[8] Keisha hoped that this kind of pressure would make them think twice before wearing the outfit again.

But as much as they worked to guard against rumors and bad reputations, the "comfortably" dressed girls in beauty school were targets for judgment by other girls across the school. The Asian hardcores, the popular white girls, the quiet Chinese girls, the alternative girls, and even some teachers saw them as "inappropriately dressed" girls who were trying to be sexy, but who really came off as "trashy." Though the preppy and comfortable styles were not that different, they represented the distinction between working-class and working-poor vocational girls and middle- and working-class academic girls. As well, the "skanky" reputation of beauty girls was often connected to the "outrageous" styles of the Hispanic girls who were (sometimes incorrectly) associated with the Aesthetician program. While not all Hispanic girls at ESH were in beauty school, and certainly not all beauty girls were Hispanic, the two were often conflated. Many of the Italian and Persian beauty girls were thought to be Hispanic as well because of a shared style, a style that Isabel proudly called "dressy." As she announced one day, "I'm a dressy girl. I like makeup. I'm a makeup fanatic."

Isabel was from a working-class family that had emigrated from Brazil when she was little. She explained that she had come to her "dressy" style through her mother's "elegant" influence. As she said: "If I look like, more like *down,* not too dressy, I feel, I don't feel good looking. You know? I don't feel like, appealing to myself." Being dressy was part of her culture. In Brazil, she explained, the *least* fashionable women were still exceedingly fashionable by Canadian standards. Isabel was a "high-heel chick" who liked to make sure her hair, makeup, and style were polished and perfect. Similarly, Rosa and Carla enjoyed "looking good." They saw themselves as "classy," "dressy," and "stylish" girls who were proud of their looks. But Rosa was aware that her style marked her as different within the symbolic economy that operated within the school. One day in drama class, she told me that I would be able to pick out her "crew" just by looking down the hallway. "We look really different," she said. Rosa was referring to how "dressy" she and her friends were, compared to the preppy girls who mainly wore jeans and velour suits. Rosa and her friends wore "outfits," including fitted slacks, heels or platform clogs, ruffled blouses or plunging V-neck sweaters, hoop earrings, and highly coiffed hair. During a game of "duck, duck, goose" in drama class one day, Rosa's style made her an easy "goose." Everyone seemed to enjoy making her run in high heels. After being selected a number of times, she finally took her shoes off, a result that seemed to please those who repeatedly targeted her.

Rosa and Carla were pretty and thin girls who wore "glitzy" and "sparkly" clothes that were tight and revealing. They also wore a lot of makeup that was described by other girls as the "Mamacita" look, with dark lip liner, thin, carefully sculpted eyebrows, dark eye shadow, mascara, and brownish lipstick. Often referred to as "hoochie," their style was, according to Dren, "total trash," including "crunchy curls, or they slick back their pony tails so tight it looks like they're going to cut off the circulation to their head." Leah, who acknowledged that her own pants were "tight," felt that they were not "necessarily even as close to being as tight as lots of other people's." When I asked who those other people were, she named the Hispanic girls,

> who have their hair all permed, and they gel it in these curls, and they're known, in my group of friends, as the "crunchy curl girls" cause their hair! Like, yeah, you look like you could break it or something. Just crunch! And they wear really like, not-there shirts and really, really tight

pants. And those kind of really annoying platform runners that are kind of in style, but not really. They're white and they've got little platforms. And also they wear those high heels.

Isabel explained that the kind of judgment leveled at her and other Hispanic girls for their "dressy" style meant that they were perceived as "sluts." Because she and her friends dressed up for school, some boys assumed they were "easy." Often ignored by Isabel and her crew, boys would "make up rumors, like, 'Oh, those girls are sluts!' and stuff, and it's like, half of us, seriously, some of us don't even have boyfriends. Some of my friends don't even have a boyfriend and never even had one."

The beauty girls' "comfortable" style, including the Hispanic girls' "dressy" style, was viewed in this negative light. Though they saw their mode of dressing as "appropriate," "classy," and "tasteful," their raced and classed locations in the school, and, particularly within the vocational program of beauty school, singled them out for derogatory labels, such as "trashy" and "hoochie." Even some teachers viewed the beauty girls in this way, as Ms. Mackenzie told me. She had heard a teacher in the staffroom suggest that all you have to do with Hispanic girls is "'wind them up and watch them walk.'" But the identification that beauty girls shared with the "comfortable" subject position was still a powerful one. As Keisha noted, the beauty girls had "an image to show." She explained, "you have to look the part, right? You can't come in sweatpants and expect to do someone's makeup or hair or facial!" For the beauty girls, then, being comfortable meant being able to dress the part for which they were training, a part that, though it did not include the prestige of wearing brand-name styles, still catered to emphasized femininity and a fashionable look. Their kind of comfort was being recognized as "good" girls, who did the "right things." But for "sporty" girls, comfort meant the opposite of what it meant to the beauty girls. It meant wearing the very sweatpants that Keisha and company would never have been caught dead in at school.

"NOT TOO FANCY": SPORTY, SKATER, AND TOM-PERSON

When I asked Diane to describe her "sporty" style, she referred to the popular Asian girls who wore JLo suits. Her "sporty" was a "*different* kind of sporty*," she said. She explained that girls did not often wear her kind of sweatpants "because a lot of girls think that if they wear like, sweatpants and stuff, people will perceive them as a slob, and they don't wanna—they want a *boyfriend!*" To offer proof of this claim, she recounted the time she was standing in the cafeteria lunch line when she overheard two preppy girls talking in front of her. One girl said to the other, "'only *slobs* wear sweatpants.'" Diane, who was wearing her sweatpants at the time, muttered under her breath, "Oh, *thank you!*" The sweatpants that Diane had on were worn regularly, along with thick socks, sport shirts with three-quarter length sleeves, and loose fitting sweatshirts. It was a different "kind" of sporty than the JLo suits, Diane suggested, because the sweatpants that marked her as different from the Asian hardcores were worn for "actual" physical activity.

Diane's identification with the sporty style was also linked to her disidentification with the girls who did not "quite fit into" their low-rise jeans. "Like, tight is one thing," she declared, "where it's just tight, but you look *nice* in it. And then you have the girls who kind of wear the tight ones that bunch them and then they kind of like, hang over a little bit on the sides and stuff." Because Diane was a plus-size girl, she would not and felt she could not wear low-rise jeans. "My clothes *fit* me!" she asserted. But by grade ten, she was used to being looked up and down by preppy girls, who "would just kind of—you know that look people give you when they look at you from head to toe? And they have that kind of condescending, *'what are you doing?'* look on their face and they like, put their eyes up, in that confrontational way." I asked Diane if she had ever addressed "that look." "No, because, I've figured out that usually that look is insecurity. Maybe they're insecure about what they're wearing and so they decide that if they try to bring me down on what I'm wearing, they won't feel so bad, perhaps?" With her words hanging in the air, Diane sighed, "It's a shot in the dark, but—"

Diane played baseball, soccer, and swam regularly. When she called herself "sporty," she meant it, as did other girls who occupied this subject position. Like other girls, however, Diane's identification with the sporty look was ambivalently predicated on her rejection by preppy girls. She started high school with a deep desire to "fit in" with the "cliquey group of kids in French Immersion," by pretending that she "liked the same things as them" and that she "enjoyed their conversation." She also tried to "dress the same as them." Though Diane wore tight jeans, little shirts and skater shoes, she was not accepted into the group and was, instead, tormented for her efforts. She also recognized that "it wasn't *me*, and I never felt comfortable wearing those clothes. Ever!" Diane's critique of the preppy girls in her social scene extended to all aspects of the preppy subject position. She found their conversations about boys, makeup, music, and magazines to be "really bad" and she hated the way they all dressed in order to be popular "with the boys." But her critique was also predicated on the exclusionary matrix that she felt had been enacted against her. She sadly told me that she had experienced "bullying" by girls, who had made her experiences of school unpleasant. "Girls are mean," she concluded. Boys did not cause her any real heartache, but girls freeze "other girls out."

Because Diane refused to wear clothes that were tight, other girls suggested that she must be insecure about her body. However Diane had a realistic understanding of mainstream marketing. "Cute" clothes could only be bought by "thin" girls, "cause you go into all the stylish shops, like *Aritzia*, which is where everybody goes right now. You go in there, and the biggest size could like, fit around my foot!" But she also resented that, as a plus-size girl, she could not be "in style." Diane felt that if she had been "thinner," she would have been able to fit the styles that would have granted her entrée into the preppy subject position. "If I had been skinny," she mused, "I would have been fine!" Though after a moment, Diane added: "Mind you, it might not be true. They might still have done that [excluded her]."

Diane's sporty subject position was ambivalently located between her critique of emphasized femininity and her desire to fit in (and into the clothes). But like the beauty girls, she was proud of her style and her ability to resist the temptation to continuously try to be "like them." Diane felt that her sporty look announced to other girls that she was happy with her body and that she did not feel the need to change. As she put it, "I'm just comfortable with what I look like now. I used to be kind of uncomfortable, but right now I feel like this is how I should be, that I feel pretty, but that I don't like, I don't think that I should be skinny or anything. But I feel pretty, the way I am."

Diane's desire to "feel pretty" was not similarly experienced by other sporty girls, who saw themselves as "tomboys" with no interest in dressing "girly" or being recognized as conventionally attractive. Adrianna was often pressured to look more "like a girl" by family and friends, who suggested that she let her hair down, wear lip gloss, put in contact lenses, and wear more "stylish" clothing. But being a skateboarder who wore baggy jeans, Nike cargo pants, Firefly hoodies, and Riptorn skater shoes enabled Adrianna to avoid all of those "fancy schmancy" styles. "They try to make me [look more like a girl], but they can't," she giggled. More to the point, Adrianna (who called herself a "tom-person") did not *want* to look "like a girl." Her identification with the sporty subject position was, for her, a way out of performing the kind of girlhood that made her uncomfortable. As a skater girl, Adrianna stood in opposition to emphasized femininity, including the styles that granted girls recognition for being conventionally sexy.[9]

Adrianna did not have to continuously reassert her power as a preppy girl who could attract boys, but instead, was granted entrée into a subject position that gave her license to perform a less feminine school identity. Unlike Diane, who (ambivalently) wanted to "fit in" with preppy girls, Adrianna desired to more fully identify with the skater subject position. She felt she was in transition from being "skater/sporty" to "full skater" and had planned to acquire more skater clothing as soon as she could afford it. While Adrianna was interested in learning how to skateboard, a sport she had only just begun to explore, she was much more interested in what the style had to offer her in terms of subject positioning within the school. It was an identification that allowed her to be recognized as someone who did not need to be "girly," an identification that provided her some relief. Not only was the expectation for emphasized femininity not a part of being a skater, but also it was diametrically opposite to the identity she was cultivating in the school. As a result, she was free to explore just how "tom-person" she could be and just how far the skater style could take her in this exploration.

The sporty subject position, like the other subject positions explored in this chapter, operated on elements of inclusion and exclusion—elements that were most easily regulated and recognized through style. While preppy girls held the most power in the school by designating which dressed bodies mattered and which ones did not, each style within the range offered its own form of inclusion and exclusion, inside and outside. Girls continuously judged each other as a way of figuring out "who" they were and where they fit in. In Valerie Hey's (1997) study of girls' friendships, she found that judgment was a significant aspect to how girls related to one another. Girls "were almost compelled to position themselves against girls who appear to be what they are not" (132). As a result, Hey concludes that girls judged others in order "to make sense of their own identity" (132). Similarly, at ESH, girls recognized other girls as the same when they did not feel the need to position themselves against each other, but were, instead, drawn to each other's subject positioning within the school.

Girls never described their own styles without also describing the styles of other girls—"preppy" girls or "skanky" girls or "hoochie" girls or "sporty" girls—in order to make clear that they were *these* kind of girls and not *those* kind of girls. Making comparisons was thus a key factor in girls' negotiations of identity through the more private process of identification. Within this process, girls also experienced ambivalence by locating themselves in subject positions that they did not always want to occupy. Here, the preppy subject position continued to hold power as the most envied social category in the school, even though girls would often invoke their hatred of preppy girls in one sentence and articulate pain at being excluded from their social groups and cliques in another. These exclusions were routine for girls, who continuously negotiated the social landscape of their school, desiring a different future, even as they were content with their present; wanting change, even as they would not change a thing. The dis/identifications that girls felt with others enabled them to create affiliations, solidarities, and tightly knit cliques. It also enabled them to further solidify who they thought they were within their school identities. Abby articulated this reciprocal process when she noted that her judgments of others were obviously revisited upon her. She was neither upset at that fact, nor concerned that she was a part of other people's disidentificatory processes. In discussing preppy styles, she understood that just as she judged others, so too would others judge her: "I look at people and go, 'Wow! That's disgusting,' you know, and I'm sure they look at me and go, 'Wow! That's disgusting.'"

As I have explored in this chapter, style was the most obvious way to indicate belonging to and distinction from a subject position at ESH. In chapter 5, I explore another highly charged facet of identity negotiation in the school by focusing on girls' uses of style in their articulation of an "image," or how they were perceived as individuals, apart from social scenes, groups, and cliques. Here, I also explore style as an expression of agency that girls used to gain control, however subtly, of not just how others saw them, but also how they saw themselves.

CHAPTER 5

"I DRESS THE WAY I FEEL": IMAGE, AGENCY, AND POWER

> If you classify yourself that means you see yourself in *that* way. I see
> myself as a million things, a million or more. I like to change things.
>   Jamie, "out-there-but-socially-acceptable" style

As I DISCUSSED IN CHAPTER 1, within a poststructural framework, there are no fixed subjects with stable and immutable identities. Identity is a product of our temporary attachments to subject positions that anchor us within specific institutional settings and discursive practices. I have suggested that girls' school identities at ESH were thus fashioned from the subject positions they were assigned within the institution of the school (gender, race, ethnicity, class, sexuality, curricular track, program reputation) *and* the subject positions that they "took up" in their negotiations of those discourses. However, it is important to note that girls' seemingly structured positionings within the school were not—even though they appeared to be—fixed and "natural." Identity categories felt automatic because they existed within larger social forces that generated systems of power, inequity, and oppression. But such forces have historical antecedents and carry social contingencies that make these "inflexible" positionings open to possible reiteration and renegotiation (Butler 1990, 1993). Identity, then, is a combination of both historically and socially contingent structural features that appear to be intractable *and* the choices we make as discursively produced subjects, who recognize how we have been positioned within these broadly based structures. The "suturing" of these two forms of positioning is what produces identity, where the contingencies inherent in each open up a space for constant negotiation.

While some feminists have argued that agency is compatible within this poststructural formulation of the subject (Butler 1990, 1992, 1993, 1995, 1997; Davies 1990, 1997; Laws and Davies 2000), for others, the dissolution of a stable self precludes the possibility for agency (Benhabib et al. 1995). According to Jones (1997, 262), we cannot be both choosing agents and discursively constituted subjects at the same time "without erring on either side." For Jones, to speak of a subject who "takes up" subject positions is to invoke a "doer behind the deed." Predicated on Hegelian teleology, this version of the subject is able to know itself through steady, linear progress and is thus able to evolve toward freedom through this knowledge. Such a progression marks the self with free will, where discourse is "the result and not the origin of human action" (Valverde 1991, 182). Seyla Benhabib (1995, 29) further argues that the "death of the autonomous, self-reflective subject, capable of acting on principle" is the death of feminism, where agency is necessary in order to carry out "the emancipatory aspirations of women." For Benhabib, the instability of the female subject "undermines the feminist commitment to women's agency and sense of selfhood" (29), making both individual choice and collective action impossible.

The "agency" debate within feminism calls to mind the agency/structure debate within sociology, where structure and agency have been conceived in binary opposition to one another, making them mutually exclusive possibilities. Within traditional sociological theory, structure is the "systematic and patterned" (Hays, 1994) limitation of a preexisting social order, characterized by a "fundamental immobility and a reassuring certitude" (Derrida 1978, 279). Structure is thus viewed as a determining factor by which experience and meaning are organized. Agency, on the other hand, has been conceived of as choice and the full consciousness of the subject that produces "individual intention" (Jones 1997, 264). Agency thus requires "consciousness, free will, and reflexivity" (Fuchs 2001, 26). In response to this binary opposition, feminist sociologists of education working within a poststructural framework have asked: "Is there a way that we can do more than acknowledge and come to understand the power that forms us and use that understanding to do something different both within and outside the confines of that forming power? Is agency possible in this way" (Laws and Davies 2000, 217)? Davies (1990, 46) advocates for an "embodied speaker," who "can move within and between discourses, can see precisely how they subject her, can use the terms of one discourse to counteract, modify, refuse or go beyond." This kind of subject, Davies (1993) suggests, is capable of forging new paths and "taking up" subject positions *within* and *because* of her discursive construction, opening up "the possibility of agency to the subject through the very act of making visible the discursive threads through which their experience of themselves as specific beings is woven" (12). In this way, subjects are capable of viewing their own constitutive processes; they can "read the texts of their 'selving'" (Davies 1997; 274).

As an ethnographer, it was impossible for me to spend time with girls in the school and not realize the extent to which they understood how they had been positioned within social, cultural, and historical forces. It was equally impossible to theorize girls as having no say in how they might re/position themselves within the school's social world. As Sherry Ortner (1998, 14) points out, by attending "ethnographically to the ways in which discourses enter into people's lives," it becomes clear that structure and agency do not exist in binary opposition, but rather that they exist in relation to one another, where discourse is "implicitly or explicitly challenged" by subjects "in the course of practices that always go beyond discursive constraints." Similarly, Bettie (2003, 54) suggests that one need not come down on the side of structure over agency or agency over structure, as both are produced within the same contingent processes: "I can have it both ways because, indeed, it is both ways." For Bettie, having it both ways means understanding how it is that girls can speak of themselves as "fixed" subjects, while simultaneously engaging in acts of agency by positioning themselves within particular subject positions that challenge or negotiate how they have been constituted within specific discourses. The contingent processes of "fixity" and "fluidity" mean that it is necessary to attend ethnographically to the ways in which girls feel "fixed" by their social and historical positionings. But it is equally important to attend to the ways in which girls articulate agency through their negotiation of these identities. In fact, separating the two seemed impossible to me as I listened to and observed girls' identity talk.

At ESH, girls spoke of themselves as fixed subjects with "core" identities, but in the same breath conceived of themselves as having the power to change "who" they were within the school's social world. Many girls acknowledged that they felt fixed by particular discourses of gender, race, ethnicity, class, and sexuality. Many also acknowledged that they felt fixed by their positionings within the school's academic and vocational streams, as well as program reputations. All of these defining features culminated in an identity that felt permanent within the school as each girl became "recognized as a certain 'kind of person', in a given context" (Gee 2000–2001, 99). But most girls did not sit idly back and accept their positionings. They worked to cultivate their sense of self by carving out subject positions from the identity materials that were made available to them. This ownership over and creative infusion of school identities enabled girls to talk about themselves as consistent and stable beings amidst paradoxical articulations of themselves as fluid and contradictory. For James Gee (2000–2001), identity is thus the culmination of all the different ways in which we have come to be known through our social performances. Gee suggests that these "strands" feel like a "core identity," because they uniquely position us within a "trajectory through 'Discursive space,'" where we are "recognized, at a time and place, one way and not another" (110). But at the same time, this uniquely personal and contextual trajectory enables subjects to "narrativize" themselves, thus making change possible through the agency of the subject.

As a result of the temporary attachments that girls felt to their school identities, one of the most ubiquitous features in my discussions with girls was a discourse of fixity signaled by declarations like Jo's, "I'm just *me*," or Diane's aspiration to be seen "first of all for who *I* am," or Maria's self-referential decree, "I see myself as *Maria*." When Abby's mom bought her "those track pants that *everybody's* wearing," she refused to put them on outside of gym class. When her mom asked why, Abby simply responded, "They're not *me!*" Keisha also invoked this language of fixity when I asked her to describe herself. "I just try to be *myself*," she said. "Like, I don't try to put a front on, or try to be fake. I am what makes myself feel comfortable."

Though girls did not commonly use the word *identity*, the word *image* was constantly invoked as a way of describing a girl's school identity. A girl's image made her "readable" within the social world of ESH, where style and image became synonymous. A girl's image symbolized her public persona, her credibility, her values, and her social standing. A girl's image was thus her most protected and cultivated possession. Given its centrality in girls' lives, Isabel had no trouble offering a definition of the term:

> Your image is the way you look. I think that sums it up pretty much, cause the way you look is your image, and that's the way you're gonna feel. Cause, if you look *yourself*, it's everything inside of you, it's how you're going to feel, it's how you're going to treat people, maybe. Like, you know? If you feel good, you're gonna treat people nice. If you don't look good, then you're just gonna be, "Leave me alone! I don't wanna talk!" you know? I think that image is *really* important.

For Isabel, as well as other girls at ESH, image and "the way you look" were synonymous, making style an extension of "everything inside of you." Shen elaborated on this correlation between style and character when she told me that girls could dress any way they wanted to, "as long as their personality *suits* it."

Discourses of fixity and fluidity were prominently featured in girls' talk about their images. The ways in which girls were discursively positioned within the school became evident when they spoke of how others saw them. As Azmera explained, image boiled down to the opinion that others had of you, an opinion that often felt carved in stone.

> Like I could sit here and fight, and say, "Oh, but this is the way I am! But this is the way I am! But this is the way I am!" But they're just going to be, "But this is what I think of you! This is what I think of you! This is what I think of you!" So, it's better just to leave it at that. And if they're willing to get to know me, then they might just learn something about me that they didn't know before.

In an effort to make consistent the way others saw them with the way they saw themselves, girls wanted their styles to appear organic, as if they grew seamlessly out of their skin. They did not want their styles to look purchased, artificial, or labored. But, significantly, when girls spoke about their images, they invoked a discourse that was not really fixed, but fluid and full of the possibility for change. While girls spoke of themselves in unwavering terms ("I'm just me!"), they simultaneously explored ways to transform their images through style in the hopes of getting others to see them differently. In this regard, style became a marker of girls' agency and their ability to reposition themselves within the discourses that constituted who they were within the school.

The overlapping discourses of fixity and fluidity that girls invoked highlighted the very real consequences of their school identities based on the temporary attachments they felt in relation to their discursive positionings, while at the same time accentuated agency in their ability to negotiate these school identities. Girls worked to transform and reinvent their images, while also working to ensure that they were perceived as "authentic" within the discursive constructions that were carved out for them. These transformations and reinventions were made possible through style as social skin. Girls used it to represent who they thought they were, who they thought they wanted to be, and how they wanted others to see them within the school's social world. And while image was decided by how others received a girl's attempts to perform a particular kind of school identity, girls had opportunities—if they took them—to perform these images differently.

In this chapter, I explore examples of girls' agency through their use of style in the production of school images, including an image to remember, an image that granted them power and authority, and an image that enabled them to maintain multiple school identities. However, it is important to note that these subheadings are not discrete categories. Girls' desire to stand out in a crowd caused them to explore multiple images; girls' desire for power caused them to "stand out." Each of these examples are connected through girls' understandings of their images as fluid, even as they were set within the seemingly fixed positionings of their school identities. This fixity anchored girls within the social world of the school (Yon 2000), but, significantly, did not keep them from experimenting with who they were and how they wanted to be seen by others. Girls understood that their images were "in play," all day, everyday, and used this knowledge to engage in subtle and/or extreme transformations—with varying degrees of success.

"NOT JUST ANOTHER NUMBER": AN IMAGE TO REMEMBER

Xiu, a self-defined "quiet Chinese girl," wanted others to see her as a leader at ESH. But when I asked her how she *thought* others saw her, she responded, "I think they think of me as a really average person." She admired all the athletic looking girls who wore Mountain Equipment Co-op Clothing, tear-away parachute pants, and "real" running shoes for track and field events. She felt that this truly "sporty" look created an image of self-control. "I want people to see like, the leader side of me, like *that* look—but then, I don't have *that* look." I asked her how she thought a leader looked. "Strong and tough!" she replied. "And how do you look?" I asked. Laughing, she said, "Uh, weak and tired!" Xiu's own style was in keeping with her "quiet Chinese" image: nice, sweet, plain, and modest. She told me that her brother-in-law often teased her about being "good," "cute," and "quiet." As a result, she wondered if she should experiment more: "be bad and rebellious, you know? But, my brother-in-law, he calls me really quiet cause I'm not like the other girls. I won't go out every Friday and Saturday, and you know, *party, party, party*." Instead, Xiu and her friends studied, did homework, and went out "every now and then. Either shopping or to the movies or just average stuff." She explained that other Asian girls, particularly hardcores, constantly went out for dinner and drank "coolers and beer and everything."

Xiu was doing well in her Mandarin class and was able to speak the language almost fluently. Because of this skill, other students in the class thought that she was from Hong Kong. She often surprised people by telling them that she was born here, making her a CBC (Canadian Born Chinese) with a strong interest in Chinese culture and traditions.

As a result of this cultural hybridity, Xiu felt that she was caught "in between the white world and the Chinese world." As she explained it, "I sort of want to be Canadian, like a Canadian, not with like, chopsticks all the time and rice! The Chinese world is, um, I guess its traditions are heavy." But then she thoughtfully clarified that she felt that the "Chinese world is winning right now [in her mind]. Cause, uh, my Chinese is starting to develop a bit more. Cause before when I spoke Chinese, it would be like, 'Oh, I can tell you're Canadian.' And now they can't tell."

Even though Xiu's language skills caused others to see her as a "Honger," she knew that she could never really be mistaken for someone from Hong Kong because of her style. Girls from Hong Kong had a different "dress code" than she did. "Underneath, they can tell that I'm a CBC," she said with self-assurance, because she did not wear the loose fitting, "criss-crossed," "strips of cloth" that many girls from Hong Kong and China wore. "And it's so different from my style, I guess." But Xiu's image as a "quiet" CBC made her feel like she was "in a box," with "tape, cardboard and everything." I asked her what the box stood for. "I think it's who I *am*," she replied. "I'm a person that doesn't really talk about their feelings. Like, I can express it with my facial expressions, but then I wouldn't talk about it until the last minute."

However fixed Xiu may have seen herself, it did not stop her from thoughtfully enacting subtle modifications to her image as she cautiously tried to negotiate how others saw her. She told me that she loved to wear the color black every now and then, because it offered "a secret, mysterious look, and that's the kind of look I'd like to keep." Because Xiu's style and personality were seen as "too kid-ish," she saw black as a way to shift this image and add maturity and sophistication to her look. Giggling, she told me that wearing black made others at school "totally get confused" about "who" she "really" was. Pausing for a moment, Xiu further commented that creating this confusion in how others perceived her brought her real satisfaction, a private pleasure that Xiu hoped to cultivate further during her high school years.

Like Xiu, Azmera also felt that she blended "in with the crowd," because, as she put it, "I look like an average person." Azmera's mother was a Muslim from Kenya, whom Azmera described as "strict."

> Yeah. We get along pretty well. We have—well, culturally females aren't supposed to be going out of the house and staying out late, and if they do, they're supposed to be with like, family members. It's just like, what they [Muslims] believe in. But being in this country, we want freedom. We want to be out with our friends and stuff.

Azmera and her mother had lived in Canada for nine years and were struggling with Azmera's desire to be a "normal" Canadian girl. Azmera was not allowed to date ("she thinks I've never had a boyfriend in my whole life!") or drink ("my favourite place to go would be *Red Robin* and *Cactus Club*"), as most of her beauty school friends did. As a result, she would often lie to her mother about where she was and what she was doing. She also struggled to break free of the traditions that her mother hoped she would take up, causing Azmera to become "more rebellious." "Because before I'd just do what she said, how she wanted me to do it and just not complain about it." But the summer before school started, Azmera found herself being less obedient. "Like, I've listened to her my whole life, so she should at least compromise. But she wants me to get into an arranged marriage and the whole thing, and I tell her 'no.'"

Though Azmera fought against her mother's strict rules and religious beliefs, she sported a style that made her mother happy. She wore no-name brand jeans, sweaters, and comfortable clothing that did not cost a lot of money. In accordance with her mother's point of view, Azmera did not agree with the "sexy" style that her friends wore to school. "The way some girls—*no!* The way some girls can dress and their parents can let them out of the house looking like that!" Azmera told me that her mom often bought her clothes that she really liked. "And it's pretty neat that she knows," Azmera said of her mom's taste. I asked Azmera if her sense of rebellion ever extended to her clothing. "No!" she emphatically responded. "Because as long as I'm covered up, then she doesn't mind. As long as I'm not showing my stomach! Cause sometimes I'll have a shirt that's smaller and it shows a little belly and she'll be like, '*Cover it! Cover it!*'"

But like Xiu, Azmera longed to subtly enhance her image as the "nice girl" who "couldn't hurt a fly"—to surprise her friends, whom she felt saw her in only one particular way. "Well, see, I like to try different things," she explained. Azmera experimented with different "looks" that she hoped would prompt others to say "'Oh! Wow! That looks kinda neat! I didn't think you could do that!'" For Azmera, standing out meant putting together "outfits" that were not "regular," such as her favorite boots, "the jean ones with the diamond sparkles on them," her light blue pants that she folded up in the 1950s retro look that was *en vogue* at the time, a shirt with sparkles on it, a sweater over top that enabled the sparkles to show around the collar, "and my hat. I love hats too." This outfit brought Azmera tremendous satisfaction; she felt that it generated "buzz" around her and enabled her to stand out in a crowd. She proudly told me that a lady on the street had complimented her on an outfit one day. "It felt great!" she said, beaming. These acts of agency helped counteract how different she felt from her friends because of her mother's religious beliefs. "I can't sleep over. I've never slept over at anyone's house," she told me.

It's just that you're not supposed to. Everything with her, it's not whether she wants me to or not want me to, it's just that she believes that you're not supposed to. She won't change. It's like, too late for her to change and learn new ways and stuff. Cause she's known just that way her whole life.

Both Xiu, a "quiet Chinese girl," and Azmera, a "nice" Muslim girl, used style to negotiate their racial and cultural identities within the school. While other girls at ESH might not have considered these changes "risky," both Xiu and Azmera were engaged in acts of agency that enabled them to gain a sense of control over the images that others expected them to have. Girls who came from strict cultural backgrounds, particularly ones like Azmera's, where girls and women had to maintain a modest demeanor, could only use style to negotiate their images in understated and quiet ways. But for girls who felt they had a little more freedom at home, cultivating an image to remember at school was much easier (and ultimately much less risky) to do.

Leah, a white, middle-class girl, felt free to experiment with her image as much as she wanted in order to distinguish herself from the girls in her preppy social scene. As a quiet popular Frenchie, Leah felt trapped by her image and desperately wanted to cultivate a style that moved her away from the Britney look, a look that she referred to as "boring and mainstream." She told me that she never really "liked" her clothes. "I just got them because that's what the stores had and that's what I thought I should have." Leah wore clothing that she and her mom picked out together at shops that she described as "mom heaven," since they catered to a "nice" and "clean" image for girls. But the summer before grade nine,[1] Leah began to think about her image and how she wanted others to see her at school. She yearned to explore a style that would foster the image that she wanted, an image that was "funky," "cool," and "unique." Leah called this style "random." The art of randomness, as Leah explained it, was being able to wear anything you wanted, no matter how "totally weird" or "clashing" or "mismatched" it was. For example, Leah aspired to eventually wear striped socks, plaid pants, and a checked shirt instead of the "typical" low-rise jeans and tight T-shirts that were ubiquitous at ESH.

In an effort to shift her image, Leah bought a pair of jeans that were "insanely studded." "I saw them there and I was like, 'Wow! These are pretty cool!'" She wore them at the beginning of the school year in order to announce her new look.

> So I wore them and probably second week of school, I'm like, "these are kind of ugly. They're tacky. Why am I wearing these?" And it was kind of after that I began to wean out of my normal style. It was just kind of like, yeah, well—buying those jeans, despite the fact that they were tacky and ugly, it was like a revolution. I'm wearing something that not everyone in Vancouver's going to have!

This foray into a new style opened the floodgates for Leah to experiment with more and more "random" articles of clothing. She began to trawl secondhand markets, like *Value Village,* where she picked up "some pretty insane things," like her florescent yellow skirt. At ESH, skirts were an anomaly in a sea of low-rise jeans and suit bottoms. As Leah recognized, "I'm sure lots of people think that it's [the skirt] kind of weird and tacky. But to me it just says a lot about *me*. It's kind of, it's in your face, but it's kind of subtle at the same time."

Leah prided herself on her "random" style and felt that she was successfully shifting her image from mainstream to "unique" within the school, though she wished she had the "guts" to "go the extra mile" and become "fully random." "I'm kind of pushing the limits a bit about what I personally feel I can pull off," she told me regarding her penchant for secondhand skirts. "But I'm also staying within that comfort zone. So to me it's kind of like I'm moving away from the normal stuff, but I'm still kind of hanging on to that last little bit of conformity." Leah did not so much move back and forth between "random" and "preppy" styles as she engaged in a hybridization of the two looks, melding elements of both together to create a style that was truly distinctive within the school. Though, others sometimes commented that Leah's efforts did not really suit who she was and seemed a little "forced." But for Leah, the "random" look was a way of taking small steps in the "right" direction; little by little, outfit by outfit, Leah planned to eventually arrive at the image she so badly wanted—a one-of-a-kind original.

Cultivating an original image was, perhaps, one of the most difficult tasks for girls at ESH, a task with which Shen was admittedly obsessed. When I first met Shen, I thought she was a "quiet Chinese girl." She rarely spoke in her classes and seemed to shy away from the bustling activity in which the beauty girls were perpetually engaged. Ms. DiAngelo, the beauty girls' customer service teacher, told me that Shen was a "quiet girl" who "kept to herself." But when Shen called out to me from across the classroom and asked if I would interview her, I wondered about the "quiet" image that she was clearly demonstrating to her teachers and classmates. I later discovered that she cultivated this image in order to deceive others into *thinking* she was quiet. She was not really interested in fostering relationships with the beauty girls and used silence as a way of avoiding the "gossip," "trash-talk," and "mainstream" mentality that she felt dominated her social scene. When I asked her about this ruse, she explained further:

> Okay, like, the beauty girls, they think I'm quiet, pretty reserved, cause I am pretty reserved, unless I want to get to know you. Cause, like, a lot of my problems with friends is that I don't really talk to everybody, just cause I don't want to. Cause if I don't really like who they are, then I don't actually make an effort to try to get to be their friend.

Shen only cared to develop relationships that were "meaningful," not "superficial," and felt that the beauty girls could not offer her the kind of conversation that she craved. "I have a few friends," she explained, "and they're SOOO close to me. Like, I'm not the type of person who likes to spread herself out, just cause they're like, deep relationships, deep friendships, right?" Shen wanted to talk about "meaningful things, not stupid things, like a lot of people talk about." She did not have a large clique within her social scene and bitterly complained that most girls at ESH were too immature for her. As a result, a lot of the beauty girls saw her as "pretty anti-social" and "shy," while she saw herself as intellectually above "most people here. Just cause I can't relate to them. They can't really seem to—like, some of the things they say out loud, they don't really seem to be that smart on the inside."

Shen had transferred into beauty school from Accelerated Studies—one of the "big three" academic programs at ESH. As a result, she saw herself as "smarter" and more "intellectual" than her classmates. As well, Shen described herself as a Christian and held values that did not coincide with those of most other beauty girls. She disapproved of how many of her classmates dressed, talked, and acted, was one of the few beauty girls who did not smoke, drink, or "party," and did not enjoy the "loose" talk that emanated from her "raunchy" classmates. These distinctions continued to separate Shen from her social scene and caused her to realize that she simply did not "fit in anywhere." She could neither hang out with her old friends in Accelerated Studies, who no longer saw her as one of them, nor could she enjoy the good-natured humor of the beauty girls, who could not figure out how to relate to her as a seemingly "quiet" Chinese girl. Though, as she continuously reminded me, Shen did not really want "more friends," she just wanted "BETTER friends"—friends with whom she could "just" be herself.

Shen described herself as someone who was not "really noticed that much, right? Just cause I'm not that outspoken or anything." She compared herself to Keisha, whom she saw as blunt, honest, fun, and wildly colorful in her personality. In contrast, Shen saw herself as living inside of her head, while hiding her "true" self from everyone around her. She admitted that she did not "blend in" with the beauty girls, "but then, I don't stand out." Thinking about this statement for a moment, she reiterated, "I *fit* in, but I *don't* blend in, and I don't want to stand out. So I guess that's how I dress." Shen's style made her a chameleon at ESH. She looked slightly preppy, slightly sporty, and slightly comfortable. She wore casual, modest, and only partially label-oriented clothing. Shen was careful to keep herself out of a "category" because "you don't want to come off as just stupid, right? Like, following media and everything like that. Following what's cool and what's now. You gotta be yourself. Everybody says that, right?" I asked Shen what she thought "being yourself" meant. "Just like, not really trying to fit in to any category, cause there's so many different categories. Like, they all revolve around looks, right? Cause that's the category."

Though Shen hated the word "cool" and all that it entailed, she also liked to look "stylish" and this desire conflicted with her working-class, Chinese background. She felt that her parents did not "really understand" her. "They have their ideas from the past and try to bring it here, and it doesn't really work." As an example, Shen cited the fact that her parents expected "respect." "Like, I see it as, if they don't respect me, I don't respect them. Like I don't just respect people just because they're older than me." Shen felt that her parents did not respect her desire to shop and look a certain way. Though her parents gave her money when she asked for it, it was not enough to buy the things she truly wanted. "I really want a job," she complained, "just so I can go, like, shopping more, cause I don't like asking them for money. Cause they say I spend too much, but compared to other people, like, I don't spend much at all!"

Shen's lack of financial freedom did not stop her from wanting to stand out from her social scene and her chameleon style enabled her to see herself as truly "unique." "First of all, like, yeah, my style? It's not really here or there. It's like pretty much everywhere." Shen felt that her look set her apart from the beauty girls. And though she did not exactly stand out in her style, she felt that it was obvious that she was different from "like, people who really follow this kind of [trendy look]." Shen avoided the Britney and JLo uniforms (with the exception of the one name brand JLo suit that she had recently bought on sale), feeling that they invited the "wrong" kind of attention from boys. Shen *did* want attention for how she looked, just not the sexual attention that made her uncomfortable. "Like, the fact that I have a *different* style, I want them to notice that. Yeah."

Wanting to be known for having a "different" style was a feeling Dren could relate to. The very mention of Dren's name in most social groups in the school caused a flurry of conversation. Teachers, guidance counselor, and students all knew Dren, if not by name, then by style. She was "that *goth* girl" to just about everyone—and she knew it. Dren's style was her "baby," something that she had fostered from grade seven on up. "I've kind of cultivated it," she said of her "look." "It's something that not a lot of people have and it makes me different and I'm proud of it cause it's my own." Dren saw her goth image as meshing with her "intense" personality. "I have strong opinions," she told me. "I got that from my mother." Dren called her mother a "feminist" who taught her how to fight for what she believed in. Dren loved to engage anyone in a discussion, but these heated debates often became fights. Most girls felt as though Dren was "crapping on their opinions" whenever she bothered to speak to them, and that she was "trying to like, shoot them down." But Dren claimed she was really "interested" in what others were saying and just wanted to provoke a "better" and "deeper" conversation.

> Like, I like to see people react and really fight for what they think. And like, all those [popular] girls [in French Immersion], they're not like that. They're like, they're just, I hate to say it, but they're the kind of people who care so much about like, social life, that they're not really into that. A couple of them are here and there, but they don't really get into like, opinionated debates like I like to. And so, a lot of the times, they think that I'm just like, being a total bitch. So, a lot of times, I don't think that a lot of them really, really like me.

Dren's left-leaning, intellectual upbringing and her white, middle-class background gave her the freedom to experiment with her style—a freedom that most girls at ESH did not have due to financial and cultural restrictions. Shitar, for example, explained that while she and Dren were both trying to stand out and that they had "that part in common," Dren had "a little more freedom" than Shitar did. "I don't wanna get kicked out of my house," she said. "Like, if I wore some of the things [Dren] wears, if I like, pierced my lip or something—which I probably will later, but not now—if I did, like, I'd get kicked out of my house." Shitar, a working-class Lebanese girl, came from a strict Christian family who, like Azmera's mother, monitored her activities and worried that she was going to be negatively influenced by other students who did not share her family's beliefs. During one of our conversations, Shitar recounted the time her parents kicked down her door after she did not respond to their knocking. "I think they thought I snuck out or died, I don't know." She was listening to loud music on her headphones and simply did not hear them. Months later, her door had yet to be replaced.

> They want my door always to be open. They always want to know where I am. So when I go places, I call them and I tell them. But they're like, "Who are you with? Who are you with? Where are you going? What are you doing right now? What are you eating?" I'm like "Shut up! I told you I'm not going to be here! I'm not going to be home! I'm at the movies!"—But they're like, "Who are you with? Are you with any boys? Boys are bad! Blah, blah, blah!"

In sharp contrast to the surveillance Shitar felt from her parents, Dren seemed to have unlimited freedom. She and Ratch regaled me with stories from their weekends each and every Monday morning. Dren lived part-time with both of her divorced parents, neither of whom gave her trouble about either her extreme goth look or her extremely independent attitude. She "partied" into the night, went drinking in metal and punk bars (if she could get in), and had enough money at her disposal to purchase expensive goth clothing. As well, Dren's father allowed Ratch to "temporarily" move in with them because she had been kicked out of her mom's apartment. As Dren explained, Ratch was not kicked out for being "punk," but rather for "cramping" her "mom's style" by being around "too much"—a fate that would never have befallen Dren.

As a result of this freedom, Dren's style reflected an audacity that most girls at ESH simply could not pull off—an audacity that had earned her a great deal of fame at school and within the Wellington neighborhood. Proudly, she told me how wide-reaching her reputation was when a stranger at a party "recognized" her:

> And I started to talk to this, like, Native guy there, and we were talking and then, all of a sudden, he's like, 'I know you' and I've never seen this kid in my life, and he's like, "yeah, you look really familiar." He's like, "Do you go to ESH?" I'm like, "Yeah." He's like, "I know who you are. I see you around all the time." He's like, "Do I look familiar?" I'm like, "No, I've never seen you before [laughs]." So it's like stuff like that. That's happened to me before, too, people who like, recognize me.

As we sat in her father's living room, drinking tea and eating the oatmeal cookies that she had made, Dren explained why she went to the lengths that she did in order to stand out at school. "I like people looking at me and thinking, 'Wow, she's really different!' Or, 'Wow, she *looks* really different!'" To Dren, being different was more than just standing out in the present. It was a way to make certain that she would continue to stand out in the future. Her image ensured that she would be remembered; that she would never be "just another number." To Dren, being goth was much more than a subculture or a particular taste in music or a style; it was much more than an alternative affiliation. Being goth was her ticket to immortality.

[Being goth means] That I'm like, my own person and I don't look like everyone else. I don't act like everyone else and I don't follow the same herd as everyone else. Something that like, makes me different in a way that maybe I'll be remembered. I mean, maybe in a couple of years, when I leave ESH, they'll be all, "Remember that goth girl, that Dren girl? Remember that goth chick who used to walk around with her chains?" You know?

Dren wanted to fill a role at ESH that she felt was missing, the role of the "misfit," or "the kind of kids who like, hang out in the back of the school and smoke their cigarettes or whatever, and kind of glare at everyone else, and walk around the school in their chains and stuff like that." Dren saw herself as one of those "misfits," purposefully keeping herself on the edge of ESH's social world, where she and Ratch spent lunchtime in the bus shelter, smoking cigarettes, picking at their chipped black nail polish, and making fun of the preppy girls who came within their laconic radar. Though Dren easily socialized with preppy girls, was invited to their parties, and shared stories with them about getting high, drunk, and "laid," she also relished being able to stand outside of that social scene by filling the "misfit" boots at ESH. "I kind of like that," she told me. "Cause no one else is [doing it]."

Abby, too, had a "phobia" about not being "remembered" at ESH and her bad-ass-schoolgirl-business-woman style was meant to ensure that people would "look through the yearbook like 17 years later, and go, 'Oh, that was Abby! She was weird!'" Abby wanted to be remembered as an "obscure" girl, and if she could not summon up such a reputation *during* high school, then she would happily wait until *after* high school to become recognized as "ahead of her time." Abby reasoned that no one remembered "these preppy girls who all look the same and they're all conformed to what the style of the time was." Abby wanted to tap into a tradition that she had learned "from movies," where people become known for their high school personas later on in life. "What

I've learned from the movies," she explained, "is that [other students] always go like, 'Wow, that's you know, like, the goth!' or, 'she was ahead of her time!'" For Abby, being ignored in high school was not that big a deal. But the thought of being ignored after high school, once everyone would clearly be able to see her for who she really "was," was too horrible for words.

Like Dren, what made Abby the most proud of her style was that people always commented on how "original" she was. "You know, people say, like, 'Oh wow! That's like, a really original outfit' or whatever. I always get really gloating and think, *Yes!* They said I was original!'" Though, originality was not always easy to maintain, even for Abby. She told me that when she saw someone who was wearing the same boots as her at school, she "got really jealous." "Well, I used to have these crazy combat army boots," she explained.

> Like, they were really cheap like, fake leather, but they were really cool. And I came to school one day with them on, and I saw this other girl was wearing them, and I was just furious! Like, I freaked out, and I was like, "What the hell is she doing with my shoes?" These are the shoes that I wore straight for a very long time. And that really made me mad, because that was sort of my style at the time—that was last year. Yeah. Yeah, thinking about that is making me mad now!

Abby's image revolved around her desire to be "one-of-a-kind," so when the trend toward tall, striped kneesocks swept through ESH, she was devastated. When someone walked by wearing a pair of her trademark socks, she would mutter under her breath, "Damn her."

Shitar, who saw herself as "strange" and called herself an "expressionalist," did not mind seeing others dressed like a "punk ass" at ESH as much as Abby minded seeing others in her patented socks. She told me that she saw a few other girls who dressed like her "around," but they were in grades eleven and twelve, older than she was and not within her social scene. She appreciated seeing older girls who engaged in a DIY aesthetic and it somehow made her feel good about her own image, offering her validation of her style. Thinking about it for a moment, Shitar remarked: "Well, I appreciate it. As long as *everybody* doesn't show up dressed like me. Then I'll dress like *them.*" Shitar explained that if everyone suddenly started coming to school in the "punk ass" look, then she would be forced to wear, "I don't know, what *they* used to wear." "Tight jeans?" I asked. "Tight jeans! Sure, if that's what it takes! Or a paper bag!"

"Nobody Starts Shit with Us": An Image of (Sexual) Power and Authority

Connected to, but also distinct from girls' efforts to stand out in a crowd, some girls expressed agency in their use of style to generate an aura of power and authority. This kind of image often entailed creating an impression of self-esteem and confidence, not caring what others thought, and commanding a quiet form of power that did not need to be overtly emphasized. As Isabel noted, sometimes a girl's image "draws you toward them. It's like they have, like powers—'come here!' type of thing." Sydney had such powers. I came to know Sydney quite well over the year, though she was never able to find the time for a formal interview. She and her mom had recently moved into a two-bedroom apartment with her grandmother and she did not want to be unavailable if her mom needed help getting things in order and organizing the "enormous mess" that Sydney told me was "constantly taking over" the apartment. Sydney wanted to be at home as much as possible, feeling an acute sense of responsibility to her family. When I offered to interview Sydney in her apartment, she quickly responded that there was no place to sit because everything was "really disorganized."[2]

I had lunch with Sydney and her social clique once in awhile, and hung out with her in classes, where we often discussed U.S. politics, her hatred of George W. Bush, and her love of Eminem. Sydney was a working-poor, multiracial girl. She was very tall, solid, and broad in her physique. She composed and performed rap songs, was a talented creative writer, and worked hard to stay on the honor role while maintaining her responsibilities at home. I marveled at how Sydney was able to remain on the fringes of popularity, while still commanding respect from other girls, even the loud popular Frenchies. Everyone admired Sydney, including Mina, who told me that Sydney was never afraid to voice her "own opinions." Mina explained that Sydney always "does what she wants and she's not afraid to tell people how she feels about them. I really admire her for that because I can't always do that."

Sydney's quiet power meant that she had a fair bit of control over her image and how others saw her. She was perceived to be wholly authentic—the "real deal"—even though she was admittedly not trendy, fashionable, or popular. She wore a hip hop style that incorporated baggy basketball jerseys, hoodies with team logos, toques and do-rags around her long, black hair, Eminem T-shirts, baggy sweatpants, and nonlabel sneakers. Her style was tough and masculine, but not intimidating, linked to a hip hop aesthetic, but not trendy. She remained unaffected by trends, not because she was cultivating an alternative affiliation, but because she simply did not care whether or not she was deemed to "fit in" by others. And this lack of interest in the symbolic economy of style that permeated ESH made her well respected by those who wished they could be as unconcerned as she was about what others thought of them. Instead she exuded a modest pride and showed kindness to everyone, and these qualities perpetuated and enhanced her powerful image.

During one of our lunches together, she told me that she had been a "heavy child," making her "so nervous to wear *anything*. I would never wear the same thing twice in one week. Never. But now, I don't care. I'll wear the same thing every day." And she often did wear the same outfits over and over without risk of being teased—a practice that most other girls could not get away with. Keisha explained that if a girl wore the "same thing three days in a row," other girls would call her "dirty" and wonder, "why's she wearing that? She just wore it!" But these insults never fell on Sydney, who drew more power from staying above the gossip, meanness, and judgment that seemed to plague other girls' experiences of their social scenes. Sydney preferred to spend her time thinking about how the media worked, who was leading the American League in RBIs, and her family.

While Sydney commanded respect quietly by wearing what she wanted day after day, staying aloof of trends, and pressures to conform to an emphasized femininity, Gwen also quietly commanded respect from girls at ESH by purposefully cultivating a gang image. The other girls in the Aboriginal program, with whom Gwen spent alternate days, sported a hip hop style that was influenced by popular female rap artist, Missy Elliott. This style entailed baggy tear-away pants with a matching baggy jacket, a "paperboy" cap, toque, or baseball cap worn on an angle, wide sneakers with thick, undone laces, long, and heavy chains ("bling"[3]) around their necks with large dangling charms (usually the girl's initial), huge hoop earrings, and multiple rings on each hand. This look also included glittery makeup, heavy eye shadow, thick eyeliner, long manicured nails, and multiple ear piercings with the occasional tattoo. Based on this common style among the girls in the Aboriginal program, Gwen stood out as unique. She wore tear-away pants, but without the matching jacket. And instead of the white, pink, or powder blue tracksuits that the other girls sported, Gwen wore black, grey, and khaki. She also wore flared jeans, tucked in tops, and little or no makeup. But like a few of the other girls in the Aboriginal program who considered themselves to be part of east side gangs, Gwen occasionally wore a bandana tied around her forehead or wrist. I asked her if the bandana "meant anything." "Not really," she replied. "Do people come up and ask you about it?" I wondered.

"In my gym class, yes." The preppy girls who swarmed around Gwen were curious about the bandana. "They're like, 'Are you in a gang?' Just be like, 'How tough are you? You ever kicked anybody's ass?'" Though Gwen denied that the bandana had any real meaning to her as a gang emblem, she was well aware of its significance within the symbolic economy of style at ESH, where gang bandanas were known to be red, black, and blue. "I only wear the black ones," she noted. "I think red's the most dangerous one." This explanation was meant to suggest that Gwen did not need to be thought of as too tough—just tough *enough* to create the image of power and authority that she was looking to cultivate with non-Aboriginal girls.

Gwen's gang image was distinctive in the Regular program, where she told me that most of the preppy girls made her feel insecure about her style, looks, and ability to "fit in" as a "Native girl." But with the addition of the bandana, Gwen stepped into a subject position that brought her prestige. Aboriginal girls were often rumored to be involved in gangs, particularly girls in the Aboriginal program, which was viewed within ESH as a "dumping ground" for "troubled Native kids." Using this positioning to her advantage, Gwen gained power by easily accessing an image that granted her automatic authority. Interestingly, Gwen never told anyone she was actually *in* a gang. She simply adorned herself with an element of style that made it possible for others to slot her into their preconceived understanding of Aboriginal girls. Rather than *re*positioning herself within the constituting narratives that structured her existence in the school, Gwen simply "worked" them, using racist perceptions to her advantage.

Gwen's image as a "tough" girl was quietly perpetuated by her use of style. But for Ratch, there was nothing quiet about her desire to appear intimidating. "I know that when I walk down the street," she told me, "people look away. They try not to make eye contact with me. They kind of get this like, nervous look, like I'm like, carrying a knife and I'm going to chop off their head or something!" Ratch's punk style and tall, broad physique granted her the kind of confidence that very few girls expressed. She would walk down the hall and call out to whomever she pleased, making fun of the "Nammer" boys, whom she called "gangstas," for their "bling-bling." "I enjoy making fun of them, sometimes to their faces," she beamed, "and I *don't* get very nice looks from them." Ratch's bravery was acquired through her love of punk music and punk style. Before she "found" punk in grade eight, she was a shy girl. But once she developed her intimidating style, she felt that she could say and do anything without fear of reprisals, both at school and at home. "Yeah, I'm pretty comfortable with myself," she boasted, knowing how uncommon this feeling was for girls, who always, it seemed to Ratch, wanted to change something about themselves because they felt "insecure."

Dressing punk created an aura of power and authority around Ratch that she wore like a badge of honor. "I used to be scared of doing lots of things," she said of her pre-punk days. "Like, I would never think of going into a mosh pit,[4] because I might get kicked in the head or something. But now I like, go in there and like, I've had my face bashed into other people and stuff, and it doesn't phase me." Though, Ratch noted that being a girl made standing in the mosh pit at punk shows a lot easier. Her attitude and aggressive dance style landed her in a lot of trouble with men, who often swung around to hit Ratch before realizing that she was a girl. "If I was a guy, then I would've been knocked in the head," she laughed. But Ratch also faced sexism and sexual harassment in the pit, where men had put their hands down her pants and tried to feel her breasts. Men have also challenged her right to be in the mosh pit at all. "One guy," she explained, "came up and he was like, 'Get out of the fucking pit!' or whatever, and I'm like, 'Noooo! I'm having a good time! Leave me alone!'" Ratch knew that if she ever got into any real trouble, other men would rush to her rescue. "But, I don't want to be dependent on anybody else except myself," she realized. "I've learned to stand up for myself, too, dressing like this."

Ratch's style had, in fact, given her a sense of invincibility that even shocked *her* from time to time. She stood up to a mugger who demanded money from her friends outside of a convenience store on Wellington Ave. When she gave the mugger a dirty look, he called her a "bitch" and advanced threateningly toward her. "I like, looked away for a minute and then I was like, 'What the hell? Why am I looking away?' So I'm like, 'You can't talk to me like that!' And he's like, 'What?' I'm like, 'You can't talk to women like that!'" The mugger continued to threaten her and "he came like, right in my face, and he's like, 'you wanna say that to my face, bitch? You wanna be slapped?' And I'm like, *'I will kick your ass!'*" Ratch's complete confidence that she could carry out this threat, combined with her intimidating style, caused the mugger to retreat. "That was pretty traumatizing," she admitted, then adding, "I'm pretty proud of myself for that one."

Ratch's image was louder than that of Gwen or Sydney, who enjoyed the quiet respect that they achieved through their styles. For Ratch, power was in your face. She liked the menacing image she fostered with her frightening presence and style. "I don't get hassled at all," she chuckled, "like, people see me as looking dangerous even though, I'm not a dangerous person. I love people, so I wouldn't hurt anyone unless they—of course, if somebody tried to hurt me, then I know that I could do some damage, but I try not to." I asked Ratch is she enjoyed the power she had to intimidate others as a punk. "I love it," she declared. "I don't dress like this because of that, self-defence and whatever, but it does take a lot of pressure off me, cause I'm not going to get mugged like, most likely. A lot of my friends who dress like everyone else, or whatever, get mugged like, get hassled all the time." A moment later she added, "I could be looking at someone in a really mean way, and they wouldn't say anything to me, but I'm not going to hurt them unless they bother me." Ratch's image, however, was predicated on the fact that the other person did not know that she would "not kick their ass" without provocation. Instead, she let her style do the talking.

An image such as Ratch's was not the only way to cultivate an image of power and authority at ESH. The most common kind of power that girls generated in the school was also the most controversial: a sexual power predicated on the sexy styles of preppy girls. Given the "close interconnections among clothing, femininity, and sexuality" (Gleeson and Frith 2004, 112), the feminist debate over sex/sexuality *as* power has been a mainstay in second and third wave discourses. As Nan Bauer Maglin and Donna Perry (1996, xv) note, "sexuality has been a contentious issue for feminists" since the 1960s. During the second wave, feminist flags were planted in "pro-sex" and "anti-sex" camps as a way of signaling an acceptance or rejection of power derived from the body. Pro-sex feminists saw sex/sexuality as a source of empowerment, where women should feel free to have sex with as many different partners as they liked, express desire, and enjoy ownership over their bodies, including when and how to showcase them in public. Conversely, "anti-sex" feminists saw sex/sexuality as a form of heterosexist control over the female body and sought to eradicate it as part of the second wave agenda. It was not that sex was considered *de facto* bad, but rather that feminists in this camp highlighted the impossibility of separating the body from capitalist and patriarchal forces that endlessly oppressed and commodified women. How, they wondered, could anyone tease out the difference between sexual agency and sexual domination given the prevalence of the latter?

In third wave feminism,[5] sex/sexuality and the body become central tenets that offer women the tools for independence, liberation, and power. This newfound bodily freedom, or what Amy Wilkins (2004, 329) terms "emancipated sexuality," reflects for many the postfeminist turn,[6] where women are seen as deriving power from individualism, instead of collective political action (Harris, 2004a). For McRobbie (2004, 6), this new standard for female power suggests that women's freedom no longer requires "any new, fresh political understanding," but is instead predicated upon an always-already form of empowerment that is "unreliant on any past struggle." The emphasis on individual forms of pleasure and sex/sexuality have caused others to label this aspect of third wave feminism "do-me" or "babe" feminism (Quindlen 1996, 3–4), where young women "use active sexuality to stake out gender independence" (Wilkins 2004, 329).

This understanding of agency through sex/sexuality is intricately bound up in the male gaze. As Wilkins (2004, 332) suggests, "the centrality of sexuality to sexism makes the task of determining women's sexual agency complex indeed." This point is particularly true in relation to the sexual agency of teenage girls. Deborah Tolman and Tracey Higgins (1994, 206) highlight the discourses that surround girls' sexuality *when* it is discussed: "good girls are not sexual; girls who are sexual are either (1) bad girls, if they have been active, desiring agents or (2) good girls who have been victimized by boys' raging hormones." According to Fine (1988, 33), not only is the "naming of desire, pleasure, or sexual entitlement" absent from "the formal agenda of public schooling on sexuality," but it is also absent from adult/media understandings of girls' style in the school. While it is difficult—even impossible—to separate sexual agency and sexual oppression, the former does not even register as a possibility in the school, while the latter is given attention as a "problem" in dire need of solving.

As I noted in chapter 2, girls who dress in sexy styles are deemed to be in trouble or out of control by psychologists for engaging in behaviors that are "out of step" with a "normal" development. But as Kate Gleeson and Hannah Frith (2004, 103) suggest, "clothing might be one of the few opportunities for young women to explore and publicly present their sexuality." In their study of girls' explorations of identity through consumption, they locate a distinctly pleasurable narrative around the presentation of "mature and sexual identities" (107) in relation to style. While girls did not explicitly articulate what their items of clothing "meant" in relation to sex/sexuality, they acknowledged the gratification and power these styles brought them in the school. Certainly, this form of "girl power" has been rigorously critiqued by feminists who argue that the claim of sexual agency cannot hold if the real objective is to attract the male gaze and compete with other women for male attention (Griffin 2004; McRobbie 2004; Taft 2004). Yet as an ethnographer, it was impossible for me to discount the fact that dressing in "sexy" styles had the potential to offer girls a form of authority in the school.

To be sexy (but not skanky) was to have power at ESH, as Ms. Mackenzie pointed out. "I mean, I guess in a sense it is empowerment," she suggested, "because what they're saying is, 'I have the body and I'm going to flaunt it.'" Girls who wore the sexy styles of the preppy uniforms were often engaged in a performance of their "sexual selves" (Gleeson and Frith 2004, 105). But Ms. Mackenzie, like other teachers, did not count this kind of power as a positive image for girls and instead saw it as a form of emphasized femininity, where girls made themselves sexy within the heterosexual matrix of the school in order to attract boyfriends. "I think it's a shame that that's where they decide they have power," she remarked. Some girls also felt that the preppy styles at ESH produced a negative image for girls. Gwen saw sexual power as "gross" because it gave women "a bad name" and gave "guys the impression that *all* girls should dress that way." Importantly, Gwen understood that as an Aboriginal girl, she had to guard against charges of "sluttiness" even more so than girls from other cultural backgrounds, particularly white and Asian girls.

While the idea that dressing in a sexual manner to get power in the school was considered to be "a shame" by some teachers and girls, others saw the potential for authority in girls' sexy styles. Shitar was of two minds about it.

> If they [preppy girls] dress sexy to show off, then I don't like it, but if they do it cause they're *confident,* and they're not self-centred, but *confident,* it's good to be sexy sometimes. I admire that. But not always, you know, wearing shirts up to here and always like, grinding with guys at dances. Not that kind of thing!

Chrissie, too, believed in the power that girls and women could generate as sexual beings, as long as they were "being themselves." For Chrissie, being "yourself" meant having the right to express your sexuality by wearing sexy clothing. She used Britney Spears and Christina Aguilera as examples. Britney Spears, she theorized, was only dressing sexy to create a media image, to market herself in a particular way. But Christina Aguilera was truly interested in experimenting with her own sexuality for the sake of personal growth and experience. For Chrissie, the difference between the two entailed analyzing how many image changes they had each undergone. Britney Spears could not "make up her mind" about how she wanted others to see her, whereas Christina Aguilera seemed more "honest" about her desire to have a sexual image.

Mina also agreed that sexual power was a valid form of empowerment that girls should utilize if they could. "I'm a tomboy with a Lolita complex," she explained, referring to herself as a girl who used her "sexual dominance to her advantage." "I realize that I could get what I wanted if I had to," she boasted. Mina had recently begun experimenting with her sexual power when she wore a skirt to school one day instead of her trademark baggy jeans and men's shirts. She realized that people took notice of her more, commented on what she was wearing, and made her feel "confident" about her "look," something that did had not happened before in her usual "grungy" style. "I should be good enough for myself that I don't have to rely on these things to elicit compliments," she reasoned. "I mean, it's sort of like cheating, you know? If you have to use a skirt to get someone's attention, then obviously they're too shallow to bother paying attention to." But when I asked her why she did it then, she invoked a discourse of essentialized femininity to justify her decision: "I think it's hormonal. I think I'm supposed to [use my sexuality]. Like, it's just natural."

Mina's foray into sexual power seemed contra to her well-cultivated image as an alternative girl. But having experienced sexual power, Mina now claimed to have "two powers": one based on her "creepy" image, the other, based on her newfound understanding of a "natural" female sexuality.

> On the one hand, there's the part of me that can get people to do what I want. Like, move off the couch simply by threatening to sack them [kick them in the testicles]. And they know that I will actually do it. On the other hand, I can get people to move simply because—I once got a bus ride after I realized I had lost my bus ticket. And it was raining and I was in this skirt, and I just started crying, and the bus driver let me on for free. It's just the simple power. You can get guys to do what you want if you dress a certain way.

Mina was aware that dressing in a sexual way gave her a "simple" power that brought immediate results, an equation that other girls were well aware of. Dren noted that she and Ratch had no problem getting into *Indigo,* a bar in east Vancouver that was notorious for fighting, live music, and a tough, metal-head crowd. "I hate to say it," she told me, "but if you're a hot chick, you're going to have a way better chance of getting in." Ratch further explained that all she had to do was "slut" herself "up" and "put on some makeup" in order to have free passage into any bar in the city. "Basically," she explained, "it's all about the boobs."

And preppy girls, like Chrissie, were also highly aware of the power they had over both boys *and* girls when they dressed in styles considered to be sexy. Boys, she explained, "will notice you" if you dress "sexy." From her own experiences with friends, she also knew that girls would envy her and covet her style, though this power backfired on Chrissie when some of her friends decided they did not want her "around" any more because she dominated the attention of boys. As I discussed in chapter 4, the real power that preppy girls sustained was the power to regulate femininity within the school, where they dictated which dressed bodies mattered and which ones did not.

"Different Little Identities": An Image of Multiplicity

I noticed Zeni on Election Day early in September. She was wearing a bright yellow tie around her crisp white blouse. To wear a tie at that time was to risk being labeled an Avril wannabe.[7] But Zeni did not concern herself with such insults. She knew that she would never dress like a pop star that she deemed to be a "poser" for her "fake" connections to skateboarding and punk. "Oh, she thinks she's *all* rock! Look at her music! It's not rock either," Zeni mused. "It's pretty poppy—so is her face!" Instead, Zeni had worn the tie in order to make a statement on Election Day. "It had a professional feel. And, you know, I'm thinkin' 'Aw, if I'm gonna lose, I'm gonna lose in style! I'll lose in my neon yellow tie!'" It was a "scary" look for Zeni, who was concerned that the tie would send the message to others that she was "overconfident" and that she knew she was going to win. But the message that Zeni hoped others would read in the tie was her dedication and commitment to getting the job of student representative for grade ten done, and done right. "I think I was trying to send a message that I don't really care what happens today. I'm still going to find a way to get my voice heard. So even though I'm not going to get elected, I'm going to go to student council, tell them what I want done and, yeah."

When I noted how amazing it was that you could say so much with a tie, Zeni nodded slowly. Everything Zeni wore seemed to "say" something to teachers and students. "I'll wear leather pants one day and everybody's like, 'Woooo!' You know?" This reaction was multiplied when Zeni was seen carrying her electric guitar to band class. "So they say, 'Wooa! The whole rock star get-up today!'" In fact, no matter what Zeni wore, it felt to her as though the whole school was watching and waiting to hear "the story." This scrutiny made Zeni "scared to dress up." Any change that she made to her "sporty" and "casual" style became a "big deal." As she explained it, "I do a little thing and people seem to notice every little change that I make, and they just pin me to it. I don't know why they see me that way."

By late September, Zeni was the student representative for her grade, a star on both the girls' volleyball and basketball teams, and was making As, just as she always had, in her classes. Zeni also played guitar in a band that performed on talent night and was known as a capable songwriter.

Many students recognized her; when Zeni walked down the hall, she said "hi" to almost everyone she passed. But her popularity was bittersweet. Being so well known also meant that others scrutinized her and discussed her every move, and every outfit. Zeni was often teased for being "rich" in a school where most girls could not afford to own several guitars, as she did. "I just save my money and buy good things," she told me. Though she worked hard to achieve her place of respect on sports teams, on student council, and in classrooms, Zeni was trapped in an image that she felt was "unfair." Everyone thought she was "Miss Perfect." "And I don't really like it that much," she added. "It seemed like a joke at first, but they keep repeating it over and over again—." If Zeni received less than an "A" on a test or somehow did not respond the way people expected her to in class, she knew she would be asked to explain herself in a manner that was not equally applied to other students. By her own admission, she was a perfectionist; her "lowest second goal" was "pretty high to some people." As a result, Zeni admitted that she was a little "self-conscious sometimes." Reflecting for a moment, she added, "Yeah, I don't wanna really give people that—*that* much."

As a way of mitigating the surveillance of others, Zeni actively worked to cultivate multiple images at ESH in order to "keep people guessing" about who she was.

> I like to be mysterious. Gets people going about you and makes them wanna like, get to know you. Know what I mean? There's something about you that they don't know, but they wish they could know, almost. I don't like people knowing [about me]. I keep people guessing. It's fun. Cause they never know.

In order to keep people guessing, she began diversifying herself across the school, joining new sports teams, different music groups, and floating between different cliques and social scenes. Because she was everywhere in the school, she hoped that others would think of her as unknowable and unpredictable. Proudly, she told me that people now saw her in "different ways. Like, people in my guitar class see me as, say, musical. People on my soccer team think of me as, well, soccer. Basketball, just basketball." Amazed by her own ability to cultivate multiple images, she added, "I've got all these different little identities and I play them when I get there." Rather than feeling exhausted by this diversification, Zeni felt liberated by the freedom that it brought her. "It's fun to say, 'Oh, I'm this for today! Or 'This is who I'm going to be now.' Yeah." As a result of this multiplicity, Zeni described her image as "different." "That's the proper word," she realized. "I'll talk differently with different groups of people." She then thoughtfully added, "it's just different me-s. Different ways to communicate to them."

The multiplicity that Zeni fostered necessitated an eclectic style. To maintain her "sporty" and "casual" style meant that others would continue to see her in the same way and easily "pin" her down. "Like, I got called Gap girl once, but I don't wear anything Gap," she explained, but the "insult" made her realize how her style was being perceived by others as "boring." "I wanna stand out," she admitted. "Just so people can look back and say, 'Oh, I remember Zeni!' Not just, you know, *someone.*" Since everyone noticed her every move anyway, she made the decision to "go [all] out" and really change her style into something that kept everyone guessing as to who she was. She began shopping at stores that were less popular with girls at ESH, buying "unique" articles of clothing, while still maintaining a look that incorporated cargo pants, jeans, and a "sporty" aesthetic some of the time. But other times, Zeni dressed up or dressed down. Once, she wore her pajamas to school and, on special days, she wore her leather pants, white blouse, and tie. "I have different looks everyday," she declared. And these different looks enabled her to maintain her multiple images in the school. Her hope was that the more she stood out as being "unique," the more she could control how others saw her. But she also held a contradictory desire that others would help her to find out who she was, too. The more she stood out as having multiple images, the more of Zeni she hoped people would get to know. "And then they can just tell me how I am—*who* I am. Save me all the trouble."

The irony of this statement was not lost on Zeni. She realized that she did not want others to know everything about her, while at the same time wanted others to help her learn more about herself. Zeni compared this contradictory element in her school identity to Britney Spears, who,

> worked hard to get to where she is now, and she's all popular and that's what she wanted, right? That's all she wanted was to be known, to make money, to sing her songs. And then she gets mad with everything else that comes with it. And what comes with it is popularity, the press, photographers, all that stuff. And she's getting mad about people knowing her, and wanting to know her.

Believing that she had done the same thing as Britney Spears, Zeni called herself a hypocrite who wanted and did not want attention in the school. Upon further reflection, however, Zeni reasoned that Britney was unwilling to accept the consequences of her own concocted image, while Zeni was willing to negotiate people's desire to know her with her own desire to be un/known. As she thoughtfully suggested toward the end of our discussion, "you have to be prepared for the part. It's like if you wanna commit a crime, you've gotta be prepared to, you know, do the time."

Zeni's commitment to fostering multiplicity in her school image was also felt by Gwen. As a student in the Aboriginal program, Gwen spent alternative days "upstairs" and "downstairs."[8] As a result, she was engaged in a highly intricate usage of style in order to negotiate what she perceived to be the social and cultural constraints of each discursive and physical space. Gwen dressed differently for "upstairs" and "downstairs." She explained that she felt the need to adopt a preppy style "upstairs," a requirement that was actually frowned upon by many Aboriginal girls "downstairs," who sported a much tougher, hip hop style. Though, Gwen was quick to align herself with the "comfortable" subject position rather than the preppy one by explaining that although she dressed "more like *them* [preppy girls]" when she was "upstairs," she did not dress "as *bad* as them. Just be like, like *this* [pointing to her outfit], but with like jeans and a better shirt." When Gwen suggested that she did not dress "as bad" as the preppy girls "upstairs," she was referring to the amount of skin that she felt they showed. It was important to Gwen not to be perceived as a "skank" given the fact that many students already saw Aboriginal girls as "hoochies." The "comfortable" subject position was thus a safer choice for Gwen, who wanted to fit in with white and Asian girls, but did not want to (nor could she afford to) buy the mandatory labels necessary to be preppy.

But on the days when Gwen was "downstairs," she sported an altogether different look. She described herself as a "punk" who wore baggy cargo pants, loose fitting tops, spiked wristbands and collars, windbreaker jackets, and hoodies. She described this "split personality" as "punk-slash-prep." Gwen understood that her participation in two curricular programs that were located in different areas of the school enabled her to negotiate two conflicting school identities. But much more than that, Gwen also felt that her location in two different programs necessitated two different identities that enabled her to comfortably endure each discursive and physical space. While she did not wish to be aligned with white and Asian girls, whom she saw as disrespecting Aboriginal people, she knew that it was useful to fit in with them in order to both alleviate racist sentiments and distance herself from the "crazy Natives" who were "downstairs." And while she did not wish to be aligned with some of the Aboriginal girls "downstairs," whom she felt gave her people a "bad name," she also did not wish to be identified with white or Asian girls, whom she saw as "really different" from her.

In order to maintain this complex negotiation of identity, Gwen split her closet in two. She had "a whole bunch of jeans on *this* side and a whole bunch of sporty clothes on *this* side." Sitting in front of her closet each night, Gwen carefully considered her geographic and discursive location for the next day and planned her outfit accordingly. Though, when I asked which style she preferred, Gwen emphatically responded, "PUNK!" She was more comfortable in her "downstairs" style, a style that she felt better reflected who she was. Her "upstairs" style, she reasoned, was more of an act, one that Gwen cultivated in order to gain some control over how others saw her as an Aboriginal girl in a program rumored to be full of "troubled," "dangerous," and "drug addicted" kids. For Gwen, this negotiation was more than just fashion or frivolity; it was more than just agency and image. It was a serious skill used to navigate the multiple social landscapes at ESH, some of which she deemed to be unwelcoming, hateful, even threatening at times.

These examples of agency highlight girls' uses of style in the careful and creative cultivation of an image. This cultivation entailed working within the constraints of girls' discursive positionings, sometimes using such constraints to their advantage and sometimes working to reposition themselves within these constraints as best they could. Style thus enabled girls to gain some sense of control and power over their school identities, making changes to their images possible, though not always successful. Girls spent a great deal of time thinking about how they dressed and how they wanted to look in order to be seen in ways that made sense to them, given who they thought they were. As girls' images had more to do with how others saw them than how they saw themselves, many put enormous amounts of energy into trying to make these two views mesh. If a girl was seen in a way that she did not like, she used style to try to bring her image back to the place where she wanted it to be. If others pigeonholed a girl, she used style to keep them guessing and throw certainty into doubt. If a girl felt restricted by her cultural or religious background, she used style to negotiate her image subtly and delicately. If a girl felt trapped by racist stereotypes, she used style as a shield or weapon that enabled her to maneuver in and out of uncomfortable or painful situations. If a girl wanted to flirt with the idea of sexual power, she used style to try it on and test it out. In these various ways, girls enacted agency in the social world of the school and, though they did not always achieve the results they were hoping for, could continuously attempt to make change in their styles and, thus, and in their lives.

CHAPTER 6

DRESSING THE PART: DEEP SURFACES AND CONTINGENT CONCLUSIONS

> I don't think you can come to any final conclusions. It's just—it's constantly changing. There's no one statement you can make about a group. You can make generalizations, but that's about it.
>
> Mina, "grungy" style

DRESSING THE PART, the subtitle of this book, comes from a conversation that I had with Zeni one dreary Friday afternoon while we were waiting for Ms. Mackenzie to open the door for English class. We were discussing the latest Britney Spears controversy relating to her video, *(I'm a) Slave 4 U*. In the video, Britney dances in a large, warehouse-like sauna to a heavy electronic beat with a mass of orgiastic revelers. When the video begins, we see Britney in skintight, low-rise jeans and a flesh-colored top that has been cropped just under her breasts. Her hair is a wild and tangled blonde mane. Her lips are heavily glossed and gently parted. Later, we see her in a hot pink bikini top and a matching pair of "crotchless" underwear worn *over* her jeans. She is dripping in sweat, having been whipped into a frenzy by the music and the orgy-like dance. Like all of Britney's carefully designed videos, the effect of *(I'm a) Slave 4 U* is startlingly sexual. However, given this obvious intention, Zeni wondered why, in recent interviews, Britney had complained so vehemently about the way the press was criticizing her style. Zeni could not understand her grievance. Wasn't she dressing provocatively *on purpose?* Wasn't she *trying* to draw attention to herself? Given these facts, Zeni wondered why Britney did not simply "own" her style instead of complaining about the negative attention she was receiving. After all, she had made her proverbial bed. Why, then, was she not content to lie in it? As Ms. Mackenzie opened her classroom door and we began to slowly file in, Zeni turned to me with one last remark on the topic: "It's like if you're going to dress the part, you know, you might as well *play* it."

The phrase stuck with me. What was the connection between dressing for and playing a part at ESH? While style offered girls the chance to dress for parts they wanted to play as a creative form of identity construction and negotiation (preppy, punk, goth, DIY, classy, respectable, athletic, gang, hip hop), it also constrained girls within identity categories that felt nonnegotiable in the school ("quiet Chinese girl," "large" girl, beauty school girl, Hispanic girl, Aboriginal girl, working-poor girl). Dressing the part at ESH meant understanding the doubleness of identity. Girls had agency and power over the construction of their images, but others also positioned them through identity categories, curricular streams, and internal and external reputations of the school. Girls with the most power to negotiate how others saw them were the ones who felt the least constrained within the school's social world. Though, even preppy girls who could afford (and had the "right" bodies for) the Britney and JLo uniforms still expressed a certain lack of freedom to negotiate how others saw them. Quiet popular girls like Leah felt that, to some extent, they were "forced" to perform girlhood in particular ways that they did not like. They felt "trapped" in a narrow definition of "hotness" and popularity that meant they could not explore alternative aspects to their school identities. Afraid of the consequences of any real change, Leah struggled to move beyond social expectations and the need to be seen as popular, while still working to gain approval from other preppy girls, whose respect she desperately wanted.

Conversely, girls with the least amount of power to negotiate how others saw them were the ones who felt the most constrained in the school: beauty school girls, Aboriginal and Hispanic girls, working-poor girls, and girls who were considered to be "fat." But girls who occupied these subject positions were not unaware of these positionings and used their knowledge to negotiate school identities within their own social scenes, groups, and cliques. While the preppy subject position may have been out of reach to some, given its adherence to a particular standard of heteronormative femininity and label-worship, other negotiations were possible through a girl's deep understanding of how the school worked and how they were situated within its vast social world. Further, preppiness was not off-limits to working-class girls, offering the possibility for more social mobility than schools with linear social hierarchies and homogeneous populations. Girls at ESH knew that their school was "not like the movies" in this regard. While there were hierarchies across racial groups and school programs, no one group of (white, middle-class) girls "ruled the school." While not just anyone could be popular, at ESH, multiple social hierarchies made it possible for girls to achieve—if not supreme popularity—some amount of social recognition and comfort within the school.

This book has sought to highlight these complex and strategic nego-tiations of identity as they were made manifest through style. At ESH, style acted as a powerful and shifting set of signifiers that enabled girls to dis/identify with each other, recognize each other as the same or differ-ent, build solidarities, groups, and cliques, create distance and division, embody ambivalences and fluid subjectivities, forge images, invoke agency and power, and gain some measure of control over how others saw them and how they saw themselves. In this regard, dressing for school was no simple event, but entailed an understanding of all of these contingent processes, particularly the potential to be positioned within the abject (though shifting) subject position of "skank." Yet no matter how impor-tant these processes might be in girls' lives, they remain largely margin-alized through moral panic in the press, as well as academic, professional, and commonsensical discourses that generate single certainties about girls. The stories I have explored here highlight just how central style is to girls in the school, no matter what subject positions they occupy or images they seek to cultivate. For all girls, style is an important consid-eration, even if that consideration is a dismissal or a refusal to partici-pate. "No style" is still a style in the school's symbolic economy. Refusing to care about how one dresses is still an aesthetic in need of cultivation in some regard. The "just rolled out of bed" look or the "picked it up off the floor" look or the "found it in a dumpster" look still carries significations and, thus, implications for identity in the school.

As I noted in chapter 5, Sydney wore the same thing just about every day, but was still admired for her "who gives a shit" attitude. Deemed to be utterly authentic in her refusal to care (as opposed to an inauthentic poser), Sydney somehow transgressed the pressure to conform to the "cultural ideal" that Leah noted—and not without some sadness—was impossible to ignore, like a tattoo in your brain that you can't get off. However, Keisha, a beauty girl who cared a great deal about looking "classy" and "appropriate" for her age, felt that wearing the same thing everyday was the kiss of death. For her, there would be whispers of "dirtiness" and "laziness." The fact that she cared about those whispers (unlike Sydney) automatically meant that she could not "get away with" such a look. But Keisha would never have wanted to anyway. Her fear of being perceived as a "slutty" girl who might be persuaded to do "bad" things was bound up in her positioning as an African Canadian, working-poor, vocational student in beauty school, or as a "woman without class" (Bettie 2003). While Sydney was a working-poor, multiracial girl, her positioning as a "smart" student in the Regular program enabled her to negotiate her school identity differently than Keisha. Looking "dirty" was a risk that Keisha was unwilling to take, feeling safer in the comfortable subject position that linked her to a critique of preppy girls, while still ambivalently emulating their style.

Given the complexities of these negotiations, it was not just knowledge of ESH's symbolic economy of style that mattered, but knowledge of how one was positioned by others as a certain "kind" of (raced, classed, schooled) girl. It was not enough to simply dress for the part one wanted to play. Girls had to understand the limits of their performances of girlhood and work within those restrictions if they wanted others to take them seriously. Wearing Miss Sixty or Mavi jeans did not automatically make a girl popular, even though it signified an affiliation with the preppy subject position. Styles were not instantaneous transformations of identity. Girls had to cautiously deliberate how they could make changes to the way others saw them in order to ensure that it appeared organic to how they were already being perceived. In this regard, style was not a "free-for-all," where a girl could simply choose to move in and out of identity categories. Dressing the part was only half the story; the other half thus entailed "authentically" living the characteristics that were associated with particular subject positions. As social skin, style had the potential to generate diverse school identities for girls at ESH that, as Shen noted, still had to "make sense" given who a girl was.

The styles that girls wore at ESH were indicative of a particular time and place in history. As I explored in chapter 3, this study took place in the Pacific Northwest, in an urban Canadian city known to be laid-back and highly multicultural, and also burdened with crime and poverty stemming from systemic racism, an influx of hard drugs, and rigid class divides. ESH was saddled with a "bad" reputation as an "inner city" school, and the girls at ESH were often seen by outsiders as "east side girls" who were not only "poor," "tough," and less equipped than their west side counterparts for middle-class futures, but also "skanky" based on the association between working-class girls and sexual promiscuity (see Eder et al. 1995; Lees 1986; Lesko 1988). Interestingly, not all girls at ESH were working-class or working-poor; there was a middle-class population as well. But the school's geographic positioning within the city erased much of the class differences that existed among its student body, establishing this external discourse as a powerful part of the school's reputation, as well as an influential force on girls' own negotiations of identity.

Inside the school, however, an entirely different reputation bubbled up in the face of ESH's "underdog" status. Girls expressed pride in their east side upbringing and saw themselves as having more *savoir-faire* than the sheltered, "rich" girls on the west side of the city. ESH students loved their school and saw it as offering them ample opportunities to be "different" and to participate in subject positions that were not necessarily evident (to their minds) in west side schools. At ESH, many girls felt that it was possible to "find your own people" and to be whoever you wanted to be, even as stories of exclusion and frustration belied this autonomy. These rhetorical performances highlight how schools and identities are "made" in relation to one another, through internal and external discourses that both institutionalize and destabilize school and self simultaneously.

While certain elements of style might be said to exist in all schools across North America (preppy and alterative styles or jeans and sneakers), the significance of style lies not in how others (adults, outsiders, researchers) view it, but in its contextualized signification within the school. At ESH, girls positioned themselves and were positioned in preppy, alternative, comfortable, dressy, and sporty subject positions. But these social roles relied on each other for mutual definition and shape. The preppy subject position infused all others with ambivalent tension. Girls who were not preppy often discussed their hatred *of*, as well as their desire to *be* preppy girls in the same breath. The abject subject position "skank" also invaded all other positions for girls by operating as the perpetual outside to all insides. Not surprisingly, only one girl—Chrissie—noted her own positioning as a "skank," a label that she felt was inaccurate given her own reading of ESH's symbolic economy of style. "Skank" was a powerful category that was most often applied by preppy girls to both nonpreppy and other preppy girls from different racial backgrounds. The term was used to patrol the borders of popular cliques, racially and ethnically organized groups, and school streams. But as I noted in chapter 4, there was no universal understanding of who a "skank" was at ESH. Popular white girls thought that popular Asian girls were "skanks"; quiet populars thought that loud populars were "skanks"; hardcore Chinese girls thought that preppy white girls were "skanks"; and Hispanic girls and beauty girls were particularly susceptible to the label because of their perceived (and actual) vocational positioning within the school's curricular streams, highlighting the powerful intersection of gender and class in the construction of girls' identities (Bettie 2003; Walkerdine et al. 2001).

Style enabled girls to make visible their subjectivities and to showcase who they were and how they wanted to be seen. But style also enabled oppression, meanness, judgment, and constraining discourses of gender, race, ethnicity, class, and sexuality to filter into girls' lives. These constraints were further emphasized by girls' understandings of what they could and could not negotiate within their school identities, such as body size, the ability to afford the "right" labels, and the power others had to misrecognize them as other than who they thought they were. Chrissie felt that she was misrecognized as a "skank" for desiring to exhibit her power as a sexual being; Isabel felt that she was misrecognized as a "hoochie" for desiring to dress like a "classy" Brazilian woman; Xiu felt that she was misrecognized as a "quiet Chinese girl" even though she was sometimes bursting to get out of her "box"; Zeni felt that she was misrecognized as a "Gap" girl; Gianna felt that she was misrecognized as a "prep" because she wore a particular cut of jeans one day; and Shitar would certainly have felt misrecognized as a "class-A twinkie."

While style enabled these misrecognitions, it also enabled girls' meaningful readings of themselves as embodied subjects. Girls could (attempt to) mitigate, neutralize, or complicate their seemingly fixed subject positions. Chrissie adopted a baggy, skater style in order to negotiate her "skanky" reputation, but also maintained her Christian Aguilera look in order to maintain the power she experienced through sexual display; Dren distinguished herself from preppy girls by cultivating a hybrid goth/metal look that granted her ultimate fame within the school and beyond; Leah wore "crazy" jeans in order to slowly make the shift from preppy to alternative; Zeni forged "different little identities" in order to keep people from knowing who she "really" was; Mina experimented with miniskirts to see what a little sexual attention could "do" for her; Azmera wore carefully put-together outfits in order to showcase a side of her personality that was not consistent with her "nice Muslim girl" image; Gwen used both punk and prep styles to negotiate the "hostile" space between Asian and Aboriginal worlds, occasionally wearing a bandana—a known gang insignia—to subtly gain power "upstairs"; Shen hated "cool" people who wore JLo suits, but happily bought one for herself when it was on sale; and Abby, unable to afford or fit into preppy styles, created herself anew by wearing obscure articles of clothing that she believed no other girl in the school would have.

To an "outsider," these expressions of self may make little sense. How could a pair of jeans, cut and "distressed" in a particular way and bought in a particular store make any difference to a girl's life? How could the wearing of a certain color or hat or bandana matter? But when placed within the context of the school, girls' style takes on meanings that a casual observer will inevitably miss. While adults may think they know what girls are up to, reading "truth" from their bodies is an act of decontextualization that can only produce single certainties. When I approached the girl in the coffee shop that day to tell her that her "ass crack" was showing, she could only giggle at me. How could I know what she was trying to accomplish with her style? How could I have possibly understood the social world in which she was ensconced or the social significance of her sartorial choices? While I cannot say with any certainty what her style meant, one thing *is* certain: She was not dressed for me. She was dressed for a world that she understood, a world that included the social scenes, groups, and cliques within which she was positioned and had positioned herself. And while it was not hard to register an opinion of her look in my mind—to judge her according to my own sense of style, taste, and feminist politics—I still could not presume to *know* what that look was all about.

SINGLE CERTAINTIES REVISITED

So, what *is* it about? If there can be no final pronouncements, as Mina's epigraph for this conclusion succinctly states, then what purpose can this work serve? For me, this book highlights how girls come to be defined, described, and, subsequently condensed into single certainties that make it seem possible to know who a girl is just by looking. But this book has also been about highlighting how this kind of simplistic reduction is an inadequate way of exploring girls' experiences. I have endeavored to resist reductionism by retaining the complex nature of girls' identity practices and negotiations in the school. It is those very complexities that single certainties inevitably work against, flattening girls' identities and emptying girls' experiences of meaning. Style has provided me with not only a showcase for girls' negotiations of identity in the school, but also with an example of the complexity of those negotiations. I chose to focus on style because it was at the center of a discursive explosion, and the nature of that explosion intrigued me. Why girls instead of boys, whose baggy pants fell well below the waistband of their boxers during the same time that girls were revealing the whale tales of their thongs? Why this particular moment, at the turn of the twenty-first century when North American girls were finally receiving academic, professional, popular, and (most notably absent) feminist attention for the first time in history? But most importantly, why deny girls nuanced and intricate identities, where contradictory subject positions and ambivalent tensions might pave the way for deeper and more interesting understandings of girls?

If single certainties are what I have tried to resist in this book, then generative possibilities are what I have hoped to demonstrate. Through the lens of generativity, I aimed to contribute to a body of work that has, as its goal, an expansion of the limiting boundaries that currently surround girlhood. The moral panic over girls' style is only one such example. During the 1990s and early 2000s, there was a proliferation of moral panics that added to the incitement to discourse that not just girls' style had become, but girlhood in general. In the 1990s, the Ophelia genre generated a view of girlhood that made it appear pathological, spreading concern about the ability of today's teenage girls to withstand the "sudden" maelstrom of peer and media pressure. Girls were framed as victims not just of external forces, but also of their own bodies and cognitive development, constructing them as little monsters and shrinking violets who were no longer in control of themselves (Pipher 1994). In the early 2000s, another panic over girls' behavior swept through the media, only this time girls were not "drowning" victims, but relationally aggressive bullies who terrorized other girls in the quest for power and popularity. Painted as egomaniacs, these mean girls engaged in acts of social sabotage that seemed to overshadow the physical bullying of boys (Simmons 2002; Wiseman 2002). And in recent years, a new moral panic has spread concerning girls' sexual behavior. Girls are now viewed as too sexually permissive, pleasing older boys who offer popularity for blowjobs in school stairwells and after-school "rainbow" parties (Ruditis 2005). In this regard, girls are said to have "gone wild," unconsciously enacting a masculinized form of sexuality without reaping any of the pleasurable benefits themselves (Levy 2005; Kamen 2000).

The prevalence of moral panics over girls' "troubling" behavior in the last fifteen years has culminated in what can only be described as a backlash against girls. In 1991, Susan Faludi offered a devastating look at the backlash against women throughout the 1980s by highlighting examples of deep resentment and anger toward women at all levels of social life, including media representations, government policy, and workplace politics. In the early twenty-first century, this resentment and anxiety has migrated to girls, who have, just like women before them, succeeded in making themselves visible as social subjects. This visibility has generated a backlash that, like the one Faludi locates against women, overtly and covertly undermines girls' newly minted social power.

In her book *Future Girl*, Harris (2004b) explores the phenomenon of girls' increased visibility, highlighting "the idea that in a time of dramatic social, cultural, and political transition, young women are being constructed as a vanguard of new subjectivity" (1). From Harris' perspective, the conditions of late modernity, including deindustrialization, globalization, and the economic and social upheaval that these conditions generate, have created an entirely different world for youth to navigate (see also Bettis 1996; Walkerdine et al. 2001), not only foreclosing on old ways of being but also opening up new possibilities for identity predicated on shifting understandings of education and work. In light of these shifts, Harris suggests that girls have taken on "a special role in the production of the late modern social order and its values. They have become a focus for the construction of an ideal late modern subject who is self-making, resilient, and flexible" (6). As Harris explains, this subject is a neoliberal invention, one who is able to pull herself up by her own bootstraps and take charge of her own life—a subject that is not dependent on the state for support and, instead, finds strength through her own inner resources and social networks (see also Aapola et al. 2004; Currie et al. 2009).

Questioning why it is that girls, and not youth in general, have come to be invested with these overly optimistic possibilities, Harris points to, among other things, the influence of feminism. Through feminist critiques of education and work, girls are now the beneficiaries of more choices, more schooling, and more career opportunities than ever before. These changes to the social order have opened previously unknown avenues for girls' subjectivities as educated workers who are able to contribute to the economy, as well as to social life in general. As a result, girls have come to see themselves,

> and to be seen, as enjoying new freedoms and opportunities. They are far more at liberty to make choices and pursue lifestyles independently of their families, the state, and men in general. Young women have been encouraged to believe that "girls can do anything" and "girls are powerful." (Harris 2004b, 8)

Neoliberal rhetoric and hard-won feminist gains are further compounded by the current discourse of "girl power" circulating in consumer and popular culture (Aapola et al. 2004; Currie et al. 2009; Gonick 2006; Kelly and Pomerantz 2007), where girls are constructed as powerful social actors, unhampered by structural forces and the historical oppression of women within patriarchal culture. Currie et al. (2009) analyze girl power as a discursive formation that has been imbued with five such meanings. Through girl power, girls may understand themselves as powerful "bitches" with social clout, wild exhibitionists with unlimited sexual power, overachieving perfectionists who are exceptional at everything they attempt to do, butt-kicking babes who embody masculine strength while still remaining feminine, and unconventional rebels who resist dominant expectations of femininity through alternative lifestyles.

As Harris (2004b, 8) suggests, these influences have combined to make girlhood *the* subject of late modernity, where girls have been constructed as "flexible, adaptable, resilient, and ultimately responsible for their own ability to manage their lives successfully." These influences have also combined to make girls and girlhood the focus of a newly recognized academic discipline. Girls' studies scholars, many of whom came of age academically during the proliferation of moral panics surrounding girls in the late 1990s and early 2000s, have produced an impressive body of work on girls' cultural practices, particularly as they relate to alternative or resistant girlhoods.

These coalescing forces have made North American girls more visible than ever before as social subjects with power. But importantly, as girls become more visible in the social sphere, backlash narratives become more common, as do calls for tighter social controls and surveillance. These single certainties in the press are further compounded by research in the academy that is rooted in the dominant paradigm of positivism. The drive for proof within the social sciences continues to constrain the human experience within neat and tidy categories based on modernist claims to "truth" and "validity." But as Patti Lather (2007, 4) suggests, what we need is "a less comfortable social science," where generativity, and not certainty, is prized.

A less comfortable social science can be fostered in numerous ways that seek to push past single certainties through experimental and alternative methodologies (see Lather and Smithies 1997; Richardson 2003). The method that seemed to work best for me was a feminist poststructural approach to ethnography. In feminist poststructuralism, I found a theoretical framework that was compatible with my desire to write about girls in a way that did not reduce the complexity of their lives to simple or linear accounts, but rather made that complexity the focus of the story itself. This approach enabled me to apply a lens to research that I felt was useful in "multiplying, diversifying, and transforming the possibilities" (Gonick 2003, p. 48) for girlhood. By refusing the reductionism that is typical of positivist research, I aimed to participate in, to use Jones' (1993, 162) words, "increasing the number of ways girls can 'be.'" What researchers choose to highlight about their subjects becomes part of the taken-for-granted body of knowledge on girls. The subject positions that researchers condone or condemn, emphasize or ignore, wax poetic over or describe in the tersest of prose are what become accepted or unaccepted modes of being. But rather than representing girls as if they (and we) had unmediated access to their "true" selves, a feminist poststructural ethnographer pays attention to the inconsistencies that multiple subjectivities necessarily produce, and locates them within historical shifts and discursive processes. This kind of research strategy can make visible how girls have been silenced and marginalized, as well as how girls have been reduced to flat complexes and crises, instead of multifaceted subjects.

In applying a feminist poststructural lens to ethnography, I aimed to bring girls' voices to the fore without claiming that they were unmediated or unaffected by sociocultural forces. While I most certainly wanted to draw attention to the stories girls told about their school identities, I also wanted to contextualize them by situating those stories within the social, cultural, and historical milieu of the school, the neighborhood, and the city. I wanted to focus on how girls' style, as social skin, showed girls to be susceptible to—but not wholly determined by—external forms of power. As a membrane of permeability, style offered an example of identity negotiations set within not just the complexity of the school's social world, but also the broader sociocultural forces of global capitalism, popular culture, and professional definitions of girlhood. Feminist poststructural ethnography opened up a space for girls' voices *and* an analysis of power relations, where agency and structure are made manifest in the doubleness of girls' identity talk, as well as in my own mediated observations of girls' everyday experiences.

Single certainties deny this generativity and seek to paint girlhood with one brush and in one color. The messiness of girls' lives only gets in the way of this simple yet effective formula—a formula that I have inadvertently participated in. I received calls from a number of reporters on the topic of girls' style when I was conducting the research for this book. Would I mind doing a phone interview for an article being written on how girls are dressing "these days"? Would I mind speaking as an "expert" on girls' style? At first I was only too happy to participate, seeing these opportunities as a chance to offer my own "two-cents" and, perhaps, scuttle a few stereotypes. But after seeing how some reporters used my comments, I began to decline subsequent invitations. My take on identity as contextual and temporal was generally left aside, picked over in lieu of more provocative words that I used: "skanks," "sluts," "Britneys," and "JLos."

FASHION FORWARD

The girls I met at ESH were highly aware of style as an everyday facet of their contextual and temporal school identities. This awareness was made manifest not just in the doubleness of girls' identity talk or their understanding of identity both inside and outside of the school, but also in their talk of subject positions that they knew were looming in the future. Even as girls struggled to gain control over how others saw them in the school, they were already imagining and preparing for the images they would (have to) foster in what many called the "real world." Girls often spoke about how their style would necessarily shift when they entered the job market, got married, and/or became mothers, signaling an understanding of the discursive constraints that awaited them in womanhood.

Abby noted that she would have to change her bad-ass-schoolgirl-business-woman look when she was "trying to make a good impression on people for a job or something." Looking down at her high-heeled Mary Janes, a threadbare denim miniskirt, an oatmeal sweater that she had "liberated" from the school's lost and found box, and her trademark black and white striped tights, she added, "Yeah. Well *this* will only last *so* long." Drawing on her middle-class status and what she knew would be expected of her in the future, Leah, too, suggested that she would revert back to a "normal" style when she decided to go to university, but after she had traveled and had "millions of experiences." Shitar felt that she would change her "punk ass" look when she got married and had children. "I'll try not to be an adult," she mused, "but you can't be a forty-year-old dressed like *this*." Isabel also felt that marriage and children would necessitate a shift in her "high heel" and "dressy" style. "You're in the house," she reasoned. "Who're you going to look good for, you know? And you're *married*." And Diane told me that she would most likely change her "sporty" look when she entered "the world of work." In sharp contrast to her "*sweat* sweatpants" style, she said, "I kind of like the yuppie trendy look, like, black pants, black shoes, and then the, kind of, those fitted white shirts that have the collars and then like, a flare back or something, and then those, kind-of square glasses, like *your* glasses, Shauna!"

These understandings of future subject positions point to girls' awareness of the fluidity of identity and the shifts in style that temporary attachments to subject positions would continue to necessitate throughout their lives. But they also point to girls' awareness of future discursive constraints. At ESH, particular performances of girlhood were made possible by its decentralized and fragmented social world, and by its institutional setting. But beyond the walls of the school, symbolic economies of style would be predicated on other, though familiar, constraints as they intersected with other, though familiar, institutionalized configurations of gender, race, ethnicity, class, and sexuality. The rules, however shifting, would continue to shift again and again, creating entirely different forms of signification and ambivalence beyond high school and into what girls perceived to be their future roles as mature (married, working) woman (with children). But interestingly, instead of envisioning more freedom and autonomy in their adult existences, girls were suggesting that they would have to surrender to the mundane monotony of a "grown-up" life in capitalist society. Girlhood, for all its ambivalent tensions, seemed to be the better option, offering girls at least a modicum of freedom to creatively cultivate an image that was (to their mind) their very own.

As the girls at ESH understood only too well, school hierarchies fade and symbolic economies shift, die, and rise again. Girls dress for the parts they have been slated to play throughout their lives, negotiating identity in relation to the specificities of time, place, and their own continuously shifting subjectivities. They struggle, as do we all, to make consistent the judgments of others with their own idealized sense of self. But in telling complex stories about girls, we (researchers, reporters, adults) can open up space for new, meaningful, and ultimately generative understandings of girlhood—just as girls are already doing in their everyday lives.

NOTES

CHAPTER 1

1. In the case of Britney Spears, "infamy" refers to the overwhelmingly negative publicity that many of her career and life decisions have brought, such as her increasingly "scandalous" style, her staged kiss with Madonna at the 2003 MTV video music awards, her failed marriage to backup dancer Kevin Federline, her repeated stays in drug rehabilitation clinics, her mothering skills, her partying antics with Paris Hilton and Lindsay Lohan, and, perhaps most controversially, her decision to leave the house without any panties on.

2. Moral panics are produced when an issue is given sustained negative attention in the press, causing fear or concern among the general public. Stanley Cohen (1972) coined the term to describe the British media's vilification of "folk devils," or working-class youth subcultures, such as mods and rockers. Their street-fighting antics and unusual styles were easy to vilify, as the media set about covering a story that would claim crumbling social morals and the decay of urban society. Stuart Hall (1978) later expands the term to signify constructed social problems that inevitably lead to cries for new legislation and tighter social controls, such as gang violence and racially oriented conflicts. But as Hall shows, these cries are based on dominant society's desire to maintain power and protect their privileged way of life. These definitions suggest that moral panics highlight how the status quo is maintained through the shaping and policing of social mores. By spreading fear and anxiety about a particular group, it is virtually guaranteed that they will become the focus of extra scrutiny, surveillance, and tighter regulations that curtail their "undesirable" behavior. The incitement to discourse surrounding the "slut" look was yet another example of moral panic in the press, only this time, girls and their bodies were the center of intense examination and moral judgment, and not the racialized, working-class male youth of subcultural "gangs."

3. It is worth pointing out that the moral panic surrounding girls' style in the school is rooted in a heteronormative perspective that suggests girls' visible bodies can only distract boys and male teachers.

4. Regulations curtailing how boys dress (gang paraphernalia, rude slogans, ball caps in the school, bandanas, etc.) have long been on the books, yet they do not carry the same connotations as the prohibition on girls' bodies.

While low-slung pants that reveal boys' boxer shorts have been banned in many North American schools, given the connection between this style and hip hop music, such restrictions are more about discouraging racial and ethnic identities than sexual ones. The issue of boys' style is certainly deserving of further study. For a detailed analysis of how the new generation of dress codes constructs femininity and sexuality in the school, see Pomerantz (2007).

5. In Canada, if economic categories are discussed at all, they are discussed using the language of poverty rather than that of class. While there is no designated "poverty line," or level of income that definitively signifies who is living "in" poverty, there is a low-income cutoff (LICO) that is used to calculate "working poor" families in Canada. Being classified as working-poor is the result of a specific formula that takes into account what percentage of a family's income is spent on food, shelter, and clothing in relation to where the family lives (Statistics Canada 2001).

I use the term "working poor" if a girl's family received some form of government assistance and/or lived in government housing, or what girls commonly referred to as "the Projects." Working-poor girls most often lived in apartments with their single mothers, shared rooms with other siblings, and did not receive any financial support from their fathers, if they knew who or where they were. Their mothers earned minimum wage at shift or part-time work, such as phone solicitation or waitressing, or were sometimes unemployed and received social assistance.

Working-poor girls often described their families as "struggling" or "surviving," as opposed to working-class girls, who often saw themselves as doing "pretty well" or "just fine." The parents of working-class girls had jobs that presented more security than did the jobs of working-poor mothers, such as that of a hairdresser, flower arranger, house painter, store clerk, cook, construction worker, and delivery person. These jobs paid more than minimum wage, offered stable working hours, and sometimes included benefits through union membership. Working-poor and working-class parents did not attend university or college and sometimes did not finish high school. However, if the parents were immigrants, it was likely that they had more prestigious jobs in their home countries and were forced to start over when they were decredentialized upon entering Canada (Man 2004). The parents of middle-class girls in the study were university educated and had careers, such as that of a lawyer, professor, teacher, business owner, nurse, or computer technician.

6. I use Davies' (1990) expression "forced choices" in order to suggest that while girls enacted agency in their choices of style, they were also positioned by economic and cultural discourses that meant they could not buy or wear anything they wanted.

7. The concept of erasure (*sous rature*) is defined in Derrida's (1974) *Of Grammatology*. As Spivak (1974, xiv) writes in her introduction to the book, to place a word under erasure is "to write a word, cross it out, and then print

both word and deletion. (Since the word is inaccurate, it is crossed out. Since it is necessary, it remains legible.)"

8. These dispositions are learned through access to economic and cultural capital within the family. Economic capital is material wealth; cultural capital is the status one gains through material wealth, translating into the "knowledge, disposition, and skills that are passed from one generation to the next" (MacLeod 1995, 13). Cultural capital is the level of mastery one has in the language of high culture or in the cultural practices that are prized by those in power (Bourdieu 1984).

9. For examples, see Driscoll (1999, 2002); Harris (2001, 2004a); Inness (1998a, 1998b); Kearney (1998); Kelly et al. (2005); Mazzarella and Pecora (1999); Pomerantz et al. (2004); Wald (1998); Walkerdine (1990, 1993).

10. According to Connell (1987), emphasized femininity is when women and girls willingly subordinate themselves to hegemonic masculinity by attracting boys through heteronormative modes of sexuality and beauty. This kind of femininity is generally equated with the social reproduction of middle-class values.

11. Examples of such analyses include girls and independent zines (Harris 2001; Schilt 2003), girls and music, particularly the Riot Grrrl network of the early 1990s (Kearney 1997; Leonard 1997; Wald 1998), girls and technology (Garrison 2000; Harris 2001; Kelly et al. 2006; Mazzarella 2005; Reid-Walsh and Mitchell 2004), girls and teen magazines (Carrington and Bennett 1996; Currie 1999; Finders 1997; McRobbie 1991), girls and popular culture (Driscoll 1999, 2002; Inness 1998a, 1998b; Mazzarella and Pecora 1999; Mitchell and Reid-Walsh 2005), and fictional representations of girls (Early 2001; Havrilesky 2002; Owen 1999).

12. For examples, see Fine (1992); Fine and Weis (1998); Gore (1993); Kelly (1993, 2000); Kenway and Willis (1998); Luke and Gore (1992).

13. For exceptions, see Bettis and Adams (2003); Gonick (2003); Kelly et al. (2005); Pomerantz (2006).

14. *Degrassi: The Next Generation* (Epitome Pictures) was a popular television show on Canadian television during the time of this research. It is a spin-off of *Degrassi Junior High* (1986–1989). Both shows feature continuing story lines about teenagers in high school, but students at ESH ridiculed the "contrived" situations that the characters found themselves in, calling them unrealistic." For a critical analysis of the show and its place within Canadian popular culture, see Michele Byers' (2005) edited collection.

15. For a discussion of how naturalism is merely a continuation of positivism, rather than a full "break," see Roman and Apple (1990).

16. See Britzman (1991); Davies (1989, 1993, 2000); Lather (1991); Lather and Smithies (1997); St. Pierre and Pillow (2000); Walkerdine (1990).

17. Though most of the girls I spoke with had a very good understanding of the marketing "machine" that dominated their lives, it is also important to note that a critique of media does not make any of us impervious to its barrage of images, slogans, and seemingly urgent creation of needs. Many girls were

critical of the media, but this critique did not render them invincible to its powerful reach. Conversely, girls had power in relation to marketing as consumers with influence. I explore this power in chapter 2.

18. Many people asked if I thought of myself similarly to Drew Barrymore's character, Josie Geller, in the popular teen film *Never Been Kissed* (1999). In the film, Josie goes back to high school as an undercover reporter in order to learn about the habits of the "typical" American teenager. Going "native," she lives as one of them, including attending prom with the coolest boy in school. My response to this comparison is always to note the difference between Hollywood plotlines and university ethics boards. The former offers delightful (though implausible) romantic comedy; the latter, a binding ethical contract to "do no harm."

CHAPTER 2

1. For an analysis of this kind of question in feminist poststructural work, see St. Pierre (2000), particularly pp. 484–486.

2. Carter (1984, 207) defines consumption as covering, "a multitude of sins: symbolic readings of commodity representations, processes of sensual gratification, practices of economic and cultural exchange."

3. It is impossible to speak of girls as commodities without mentioning their sexual, physical, and economic exploitation in the global labor market, not just in developing nations, but also around the world. For further discussion, see Ige (1998) and Inness (2000).

4. Exceptions include Lewis' (1989) analysis of mainstream music videos and McRobbie's (1991) studies of popular teen magazines.

5. If industrial capitalism is characterized by the mass production and assembly line duplication of the culture industry, then a vast and all-encompassing empire of international commodity production and exchange characterizes global capitalism. As Hardt and Negri (2000, xii) write of global capital's empire, it "establishes no territorial center of power and does not rely on fixed boundaries or barriers. It is a *decentered* and *deterritorializing* apparatus of rule that progressively incorporates the entire global realm within its open, expanding frontiers." Unlike the industrialized economy of the modern era, the global economy of the postmodern era is focused on "deindustrialization—the contraction of large-scale manufacturing and the expansion of global communications, technology, and service industries" (Harris 2004b, 3). This form of capitalism moves at the speed of light, without concern for borders, nations, or cultures—taking consumer girls' culture from a mass produced industrial phenomenon to a whole new global level.

6. Popular feminism is characterized by the infusion (and, some would posit, the depoliticisation) of feminist ideals into mainstream popular cultural texts, including televisions shows, such as *Buffy the Vampire Slayer, Xena, Warrior Princess, the Powerpuff Girls, Sex and the City, Sabrina, the Teenage Witch, Veronica Mars,* and the new *Bionic Woman.*

7. McRobbie's groundbreaking study was originally published in 1977, but has since been reissued, along with her other essays, in the edited compilation, *Feminism and Youth Culture* (1991).

8. Though, as McRobbie (1991, 65) writes, her analysis of working-class girls' culture may have failed to acknowledge the realities of girls' lives by making class a more central feature than the girls themselves. She writes,

> I felt that somehow my "data" was refusing to do what I thought it should do. Being working-class meant little or nothing to these girls—but being a girl over-determined every moment. Unable to grapple with this uncomfortable fact, I made sure that, in my account anyway, class *did* count. If I had to go back and consider this problem [of working-class girls' culture] now, I would go about it in a very different fashion. I would not harbour such a monolithic notion of class, and instead I would investigate how relations of power and powerlessness permeated the girls' lives—in the context of school, authority, language, job opportunities, the family, and community and sexuality.

9. Harris (2001) revisits the idea of "bedroom culture" in the twenty-first century, naming the bedroom as the "new" space for girls' political engagement. In order to avoid the surveillance that has recently descended upon girls in the public sphere, Harris suggests that girls have "gone underground." Bedroom culture has thus shifted from Firth's (1981) understanding of romance and beauty rituals to a space where girls can enact politics through the Internet, including zines, "gURL" Web pages, and alternative music. This argument reconfigures the bedroom as a hub of power for girls.

10. For examples, see Aapola et al. 2004; Driscoll 1999; Gonick 2006; Garrison 2000; Kearney 1998; Leonard 1997; Wald 1998.

11. For a detailed analysis of postfeminism and popular culture, see Currie et al. (2009), and Kelly and Pomerantz (2007).

12. Tweens are a recently constructed demographic made up of preadolescents between the ages of nine and thirteen. Like the construction of the teenager before it, tweens are a marketing invention used to cultivate a new segment of the population as consumers. For analysis of this social category in relation to girlhood, see the edited collection by Mitchell and Reid-Walsh (2005).

13. While the title of Greenfield's book is in the singular, *Girl Culture,* I have chosen to write it in the plural form. This shift is meant to highlight the multiplicity of girls' culture and the impossibility of generalizing it to all girls.

14. The Ophelia genre is a large body of work that began prior to Pipher's extremely popular book. For other examples, see American Association of University Women (1992), Gilligan (1982), Gilligan et al. (1990), Glazier (2004), Orenstein (1994), Simmons (2002), and Wiseman (2002). While not all of these books and documentaries deal explicitly with girls' low

self-esteem in the teen years (Glazier, Simmons, and Wiseman detail a related moral panic in girlhood—relational aggression and "mean girls"), they all frame girlhood as a time of crisis.

15. For a detailed analysis of *Thirteen* as part of a discursive formation on girlhood within popular film, see Kelly and Pomerantz (2007).

16. Based on the box office and critical success of *Thirteen*, the *Oprah Show* focused on the teen years as a time of crisis for girls in several emotional episodes, including "Is your child leading a double life?" "Raising teenage girls," and "The secret lives of teenage girls."

CHAPTER 3

1. For examples, see *The Breakfast Club* (1985), *Pretty in Pink* (1986), *Some Kind of Wonderful* (1987), *Heathers* (1989), *Clueless* (1995), *Romy and Michele's High School Reunion* (1997), *Jawbreaker* (1999), *Never Been Kissed* (1999), *She's All That* (1999), *The Princess Diaries* (2001), *Thirteen* (2003), *13 Going on 30* (2004), *Confessions of a Drama Queen* (2004), and *Mean Girls* (2004).

2. I use the expressions "social scene," "group," and "clique" to denote various social organizations within the school. A social scene is the largest social unit a girl belongs to, for example, the French Immersion or Aesthetician programs. A social scene could also be a racial or ethnic grouping, such as Chinese students, Hispanic students, or ESL students. A social scene is made up of groups, a term I use similarly to Milner's (2004, 23) definition of "crowd," which he defines as "a social category, a type of subculture, a reference group, and a status group." At ESH, such examples included "quiet Chinese girls" or skaters or smokers—people who knew of each other and oriented themselves around similar cultural references, but who did not necessarily hang out in an intimate configuration. Groups are made up of various cliques that comprise girls' daily social networks and intimate friendships: "cliques are small groups that embody, transmit, and transform" (23) the larger groups in which they operate.

3. "Kappa people" was another label for ESH's largest racially organized social scene, the Nammers, an acknowledged name for Vietnamese students who were identified by their Kappa-brand tracksuits. Kappa is an Italian brand of sportswear that was hugely popular the year I was at ESH and signified automatic affiliation with the Nammer group, whether you were Vietnamese or not.

4. I am grateful to Catherine Chaput for this insight.

5. At the time of this research, Vancouver's visible minority population was 37 percent. This number was up from 31 percent in 1996 and 24 percent in 1991 (Statistics Canada 2001).

6. One of the political circumstances that sparked an immigrant explosion prior to 1997 was the return of Hong Kong by its British protectorate to communist China rule. The steady stream of immigrants out of Hong Kong during this time has been called the "Hong Kong Diaspora."

7. At the time of this research, in Canada, only Toronto's immigrant population, at 44 percent, surpassed that of Vancouver's immigrant population. New York's immigrant population was 24 percent; Los Angeles' was 31 percent (Statistics Canada 2001).

8. Even Chinatown and Little India are home to multiple races and ethnicities.

9. At the time of this study, 33 percent of east side families had an annual income of under $30,000 (British Columbia Ministry of Education 2004) and 37.5 percent of its households were classified as low income (Statistics Canada 2001).

10. In the Downtown East Side (DTES), it is common to see heroin addicts "shooting up" in the streets. While drug usage is chronic in the area, a much deeper understanding of the effects of poverty, racism, and the history of how Aboriginal peoples were and still are treated in Canada is needed to truly fathom the complexity of the issue. There have been rallying calls to "clean-up" the DTES for years, but no amount of gentrification will ameliorate the systemic traumas and violences that have been inflicted on its many disenfranchised populations.

11. While this reputation was well known within the city, I also had cause to hear west side girls discuss their views on east side girls during interviews I completed for another project between 2000–2003.

12. *Off the Wall* is a chain store that caters to teenage girls and carries expensive labels, such as Mavi and Miss Sixty.

13. When Xiu first said, "Chink" to me, I told her I found the word to be quite shocking and racist. She laughed and explained that all of her friends used the label to describe themselves, and that there were no racist connotations. She also told me that she and her friends "don't get mad about other people calling us that. It's really common now." All of the racialized nicknames in the school were accepted by the groups themselves and widely used within the school.

14. Pyke and Dang (2003) use two terms that are similar to CBC and ESL in their study of identity among 184 second generation Korean and Vietnamese young adults: FOBs (Fresh Off the Boats) and Whitewashed (those who were assimilated to North American culture).

15. Nelly is an African American hip hop artist who was constantly on top of the pop charts the year I was at ESH.

CHAPTER 4

1. Identification, as I am using it here, is derived from psychoanalytic theory, but has been pushed in new directions through poststructural theorizing of the self. Freud (1991 [1921]) originally theorized identification as a child's earliest sense of belonging through the Oedipal complex, where the mother is the ideal and the father is the rival. This emotional tug-of-war is a child's first understanding of desire for the other and takes place within the realms of inclusion (child and mother) and exclusion (father). The ego is formed

through this initial act of awareness, producing self-consciousness set within the ambivalent matrix of love and competition, trust and rivalry. As Freud suggests, identification is thus "ambivalent from the very start" (134). Lacan (1977) also theorizes ego formation through a process of developmental awareness, but one that is reflected back to the child in the "mirror stage." Identification is achieved through self-recognition, where the child comes to recognize him or herself as a being in the world. Once that self-recognition is projected outward through the image, it is then projected inward as the limit of his or her own sense of consciousness.

For Freud and Lacan, identification is the self-recognition that forms our worldview. But poststructural theory has worked to distinguish between the naturalism of psychoanalytic identification, where subjecthood is fought for and gained from within, and the contingency of discourse, where identification is viewed "as a construction, a process never completed—always 'in process'" (Hall 1996, 2). The subject is enacted and continues to be enacted through social relations (Butler 1993; Fuss 1995; Hall 1996). This "discursive interpellation of subjects to particular subject positions" (Gonick 2003, 13) can only take place within an "exclusionary matrix" (Butler 1993, 3), where the struggle for recognition in the social world is made possible by dis/identification.

2. The Freudian concept of ambivalence merges with discursive understandings of the subject to infuse self-recognition with fluidity and inconsistency. As the name for the "contradictory or mutually exclusive desires" (Flax 1990, 14) that we feel through identification, ambivalence can, as Freud (1991 [1921]) suggests, fill us with a sense of dread, cognitive dissonance, and the wish to kill the object of our ambivalent desire. But for feminists working in the area of poststructural psychoanalytic theory, ambivalence opens up more productive possibilities than those offered by a Freudian model of denial or repression (Butler 1993; Fuss 1995; Gonick 2003).

3. Maria explained *Aritzia*'s aesthetic as "urban." She noted that they "have some dressy stuff, but it's more focused on casual, laid-back clothing. And, I think some of the stuff's really original and stuff like that. I would never go into another store and see something like that kind of thing. And I think the quality's really good."

4. I say "chic" skater aesthetic because the girls who wore this style were not actually skaters. As I noted in chapter 3, the popular Frenchies hung out with and dated skater boys. The chic skater aesthetic was a variation on the Britney look that incorporated skater shoes, such as Etnies, DCs, and Vans.

5. The pro-ana movement, short for pro-anorexia, revolves around Web sites that are dedicated to the "rules" of doing anorexia "properly," such as how to feel full and how to avoid getting caught.

6. Dren's piercings were, in fact, what enabled Jamie and her to be friendly in the hallways, as they were two of the girls in the school who wore such

"hardware," forging an identification between them even though they hung out in very different social groups.

7. Punk-pop bands, such as Billy Talent, Sum 41, Blink 182, Green Day, and Simple Plan, have popularized this style, which is loosely associated with skaters and skate punks.

8. During their customer service class, I heard one beauty girl tell another that her "pubes" were showing, which meant, "pull up your pants!"

9. For a deeper analysis of the skater girl subject position, see Kelly et al. (2005) and Pomerantz et al. (2004).

CHAPTER 5

1. Most teachers and girls agreed that if a style change was going to be made, it usually happened in the summer before grade nine so that a girl could start off the year as someone "different." The going theory was that grade eight was too early in a girl's high school career to make any real image changes as girls were just starting high school and felt too nervous to experiment with their look. But by the end of the year, girls felt settled and ready to take greater chances if they wanted to try something new.

2. If a girl did not want me to come to her home, it was usually because she was embarrassed about her economic situation, so I always tried to offer alternative venues, including classrooms after school and local coffee shops.

3. Bling or bling-bling is an expression that comes from hip hop culture and means jewelry, particularly gold chains and any other "flashy" symbols of wealth and success. While girls at ESH were likely not wearing real gold, the use of fake bling still positioned them within a hip hop aesthetic.

4. A "mosh pit" is the area directly in front of the stage at a live punk or rock show, where participants can dive into the crowd and "surf" over the audience. A mosh pit generally engenders aggressive styles of dancing, such as slam dancing, where participants charge into each other and vigorously jump up and down.

5. Third wave feminism is meant to distinguish the political pursuits of today's feminists from the feminists of the second wave. However, it is important to note that third wave feminism does not see itself as a "post"feminist movement that views feminism as over and irrelevant. Third wave feminism seeks to take further second wave's agenda by continuing to focus on gender equality, violence against women, and oppression, while also focusing on issues of sex/sexuality, pleasure, technology, and the material issues facing women around the world. As Riordan (2001, 280) explains, while second wave feminism made sisterhood or sameness its founding issue, "third-wave feminism works explicitly toward understanding difference and respecting its importance to feminist thought." Third wave feminists often credit third-world feminists, critical race feminists,

and antifoundational feminists for third wave's focus on difference. The following are often seen as quintessential third wave texts: Barbara Findlen's (1995) *Listen Up: Voices From the Next Generation*, Rebecca Walker's (1995) *To be Real: Telling the Truth and Changing the Face of Feminism*, Leslie Heywood and Jennifer Drake's (1997) *Third Wave Agenda: Being Feminist, Doing Feminism*, and Jennifer Baumgardner and Amy Richards, *Manifesta: Young Women, Feminism, and the Future* (2000).

6. Postfeminism is the charge that young women and girls are "enjoying all the freedoms won for them by the women's movement without engaging in the struggle themselves" (Pomerantz et al. 2004, 547). This attitude, according to Judith Stacey (1990, 339), is "the simultaneous incorporation, revision and depoliticization of many of the central goals of second wave feminism." For a critique of postfeminism as a label that unfairly imputes a lack of political action to girls, see Pomerantz et al. (2004).

7. Avril wannabes were said to be emulating pop star Avril Lavigne, who, in the early portion of her career, always wore a tie. Though most girls agreed that her music was "okay," many did not buy her image as a "skate-punk" and thought she was a poser.

8. As I noted in chapter 3, the terms "upstairs" and "downstairs" referred to the locations of programs at ESH. The Aboriginal program was controversially located in the basement and many attributed this geographic positioning to the low status of Aboriginal students within the school.

References

Aapola, Sinikka, Marnina Gonick, and Anita Harris. 2004. *Young femininity: Girlhood, power and social change.* London: Palgrave.

Adorno, Theodor W., and Max Horkheimer. 1993 [1944]. *Dialectic of enlightenment.* New York: Continuum.

Agar, Michael. 1980. *The professional stranger: An informal introduction to ethnography.* New York: Academic Press.

Althusser, Louis. 1969. *For Marx.* London: Allen Lane.

———. 1971. *Lenin and philosophy, and other essays.* New York: Monthly Review Press.

American Association of University Women. 1992. *The AAUW report: How schools shortchange girls.* Wellesley, MA: American Association of University Women.

Ang, Ien. 1985. *Watching Dallas: Soap opera and the melodramatic imagination.* New York: Methuen.

Barrett, Michèle. 1980. *Women's oppression today: Problems in Marxist feminist analysis.* London: Nlb.

———. 1992. Words and things: Materialism and method in contemporary feminist analysis. In *Destabilizing theory: Contemporary feminist debates,* ed. M. Barrett and A. Phillips, 201–219. Cambridge, UK: Polity Press.

Barthes, Roland. 1983. *The fashion system.* Trans. M. Ward and R. Howard. New York: Hill and Wang.

Baumgardner, Jennifer, and Amy Richards. 2000. *Manifesta: Young women, feminism, and the future.* New York: Farrar, Straus and Giroux.

Bellafante, Ginia. 1998. Feminism: It's all about me! *Time.com,* http://www.time.com/time/magazine/1998/dom/980629/cover2.html.

Benhabib, Seyla. 1995. Feminism and postmodernism: An uneasy alliance. In *Feminist contentions: A philosophical exchange,* ed. S. Benhabib, J. Butler, D. Cornell, and N. Fraser, 17–34. New York: Routledge.

Benhabib, Seyla, Judith Butler, Drucilla Cornell, and Nancy Fraser, eds. 1995. *Feminist contentions: A philosophical exchange.* New York: Routledge.

Bettie, Julie. 2003. *Women without class: Girls, race, and identity.* Berkeley: University of California Press.

Bettis, Pamela J. 1996. Urban students, liminality, and the postindustrial context. *Sociology of Education* 69 (2):105–125.

Bettis, Pamela J., and Natalie G. Adams. 2003. The power of the preps and a cheerleading equity policy. *Sociology of Education* 76 (2):128–142.

————. 2005. *Geographies of girlhood: Identities in-between.* Mahwah, NJ: Lawrence Erlbaum Associates.

Bhabha, Homi K. 1994. *The location of culture.* New York: Routledge.

Bourdieu, Pierre. 1977. The economics of linguistic exchanges. *Social Science Information* 16 (6):645–668.

————. 1984. *Distinction: A social critique of the judgment of taste.* Cambridge, MA: Harvard University Press.

————. 1998. *Practical reason: On the theory of action.* Stanford, CA: Stanford University Press.

British Columbia Ministry of Education. 2004. Programs and services for parents. http://www.bced.gov.bc.ca/parent_ps.htm.

Britzman, Deborah P. 1991. *Practice makes practice: A critical study of learning to teach.* Albany: State University of New York Press.

————. 1997. The tangles of implication. *Qualitative Studies in Education* 10 (1):31–37.

————. 2000. "The question of belief": Writing poststructural ethnography. In *Working the ruins: Feminist poststructural theory and methods in education,* ed. E. A. St. Pierre and W. S. Pillow, 27–40. New York: Routledge.

Brumberg, Joan Jacobs. 1997. *The body project: An intimate history of American girls.* New York: Random House.

————. 2002. Introduction. In *Girl culture,* ed. Lauren Greenfield. San Francisco: Chronicle Books.

Butler, Judith. 1990. *Gender trouble: Feminism and the subversion of identity.* New York: Routledge.

————. 1992. Contingent foundations: Feminism and the question of "postmodernism." In *Feminists theorize the political,* ed. J. Butler and J. W. Scott, 3–21. New York: Routledge.

————. 1993. *Bodies that matter: On the discursive limits of "sex."* New York: Routledge.

————. 1995. For a careful reading. In *Feminist contentions: A philosophical exchange,* ed. S. Benhabib, J. Butler, D. Cornell, and N. Fraser, 127–143. New York: Routledge.

————. 1997. *The psychic life of power: Theories in subjection.* Stanford, CA: Stanford University Press.

Byers, Michele. 2005. *Growing up Degrassi: Television, identity and youth cultures.* Toronto: Sumach Press.

Carrington, Kerry, and Anna Bennet. 1996. "Girls' mags" and the pedagogical formation of the girl. In *Feminisms and pedagogies of everyday life,* ed. C. Luke, 147–166. Albany: State University of New York Press.

Carter, Erica. 1984. Alice in consumer wonderland. In *Gender and generation,* ed. A. McRobbie and M. Nava, 185–214. London: Macmillan.

Clifford, James. 1986. Introduction: Partial truths. In *Writing culture: The poetics and politics of ethnography,* ed. J. Clifford and G. Marcus, 1–26. Berkeley: University of California Press.

Cohen, Stanley. 1972. *Folk devils and moral panics: The creation of the mods and rockers, sociology and the modern world.* London: MacGibbon & Kee.

Connell, R. W. 1987. *Gender and power.* Stanford, CA: Stanford University Press.

Currie, Dawn H. 1999. *Girl talk: Adolescent magazines and their readers.* Toronto: University of Toronto Press.

Currie, Dawn H., Deirdre M. Kelly, and Shauna Pomerantz. Forthcoming 2009. *Girl power: Girls inventing girlhoods.* New York: Peter Lang.

Daly, Steven. 1999. Britney Spears: Inside the heart and mind (and bedroom) of American's new teen queen. *Rolling Stone,* April 15, 60–69.

Davies, Bronwyn. 1989. *Frogs and snails and feminist tales: Preschool children and gender.* Sydney: Allen & Unwin.

———. 1990. The concept of agency: A feminist poststructuralist analysis. *Social Analysis: Journal of Culture and Social Practice* 29:42–53.

———. 1993. *Shards of glass: Children reading and writing beyond gendered identities.* Cresskill, NJ: Hampton Press.

———. 1997. The subject of post-structuralism: A reply to Alison Jones. *Gender & Education* 9 (3):271–284.

———. 2000. Eclipsing the constitutive power of discourse: The writing of Janette Turner Hospital. In *Working the ruins: Feminist poststructural theory and methods in education,* ed. E. A. St. Pierre and W. S. Pillow, 179–198. New York: Routledge.

de Ras, Marion, and Mieke Lunenberg, eds. 1993. *Girls, girlhood and girls' studies in transition.* Amsterdam: Het Spinhuis.

Delaney, Paul, ed. 1994. *Vancouver: Representing the postmodern city.* Vancouver, Canada: Arsenal Pulp Press.

Derrida, Jacques. 1973. *Speech and phenomena: And other essays on Husserl's theory of signs.* Evanston, IL: Northwestern University Press.

———. 1974. *Of grammatology.* Baltimore, MD: Johns Hopkins University Press.

———. 1978. *Writing and difference.* Chicago, IL: University of Chicago Press.

Deveny, Kathleen, Raina Kelley, Jamie Reno, Karen Springen, Susannah Meadows, Anne Underwood, and Julie Scelfo. 2007. Girls gone bad? *Newsweek,* February 12.

Driscoll, Catherine. 1999. Girl culture, revenge and global capitalism: Cybergirls, riot grrls, spice girls. *Australian Feminist Studies* 14 (29):173–195.

———. 2002. *Girls: Feminine adolescence in popular culture and cultural theory.* New York: Columbia University Press.

Early, Frances H. 2001. Staking her claim: Buffy the vampire slayer as transgressive woman warrior. *Journal of Popular Culture* 35 (3):11–28.

Eckert, Penelope. 1989. *Jocks and burnouts: Social categories and identity in high school.* New York: Teachers College Columbia University.

Eder, Donna, Catherine Colleen Evans, and Stephen Parker. 1995. *School talk: Gender and adolescent culture.* New Brunswick, NJ: Rutgers University Press.

Ehrenreich, Barbara, Elizabeth Hess, and Gloria Jacobs. 1992. Beatlemania: Girls just want to have fun. In *The adoring audience: Fan culture and popular media,* ed. L. A. Lewis, 84–106. London: Routledge.

Entwistle, Joanne. 2000. *The fashioned body: Fashion, dress, and modern social theory.* Cambridge, UK: Polity Press.

Faludi, Susan. 1991. *Backlash: The undeclared war against American women.* New York: Crown Publishers.

Finders, Margaret. 1997. *Just girls: Hidden literacies and life in junior high.* New York: Teachers College Press.

Findlen, Barbara. 1995. *Listen up: Voices from the next feminist generation.* Seattle, WA: Seal Press.

Fine, Michelle. 1988. Sexuality, schooling, and adolescent females: The missing discourse of desire. *Harvard Educational Review* 58 (1):29–53.

———. 1992. *Disruptive voices: The possibilities of feminist research; Critical perspectives on women and gender.* Ann Arbor: University of Michigan Press.

———. 1994a. Dis-stance and other stances: Negotiations of power inside feminist research. In *Power and method: Political activism and educational research,* ed. A. Gitlin, 13–35. New York: Routledge.

———. 1994b. Working the hyphens: Reinventing self and other in qualitative research. In *Handbook of qualitative research,* ed. by N. K. Denzin and Y. S. Lincoln, 70–82. Thousand Oaks, CA: Sage Publications.

Fine, Michelle, and Lois Weis. 1998. *The unknown city: Lives of poor and working class young adults.* Boston: Beacon Press.

Firth, Simon. 1981. *Sound effects: Youth, leisure, and the politics of rock 'n' roll.* New York: Pantheon.

Flax, Jane. 1990. Postmodernism and gender relations in feminist theory. In *Feminism/postmodernism,* ed. L. J. Nicholson, 36–62. New York: Routledge.

Foucault, Michel. 1972. *The archaeology of knowledge.* London: Tavistock Publications.

———. 1977. *Discipline and punish: The birth of the prison.* London: A. Lane.

———. 1978. *The history of sexuality: An introduction.* New York: Vintage Books.

———. 1980. *Power/knowledge: Selected interviews and other writings, 1972–1977.* Ed. C. Gordon. New York: Pantheon Books.

Freud, Sigmund. 1991 [1921]. Group psychology and the analysis of the ego. In *Civilization, society and religion.* Vol. 12 (selected works). Harmondsworth, UK: Penguin.

Friedan, Betty. 1963. *The feminine mystique.* New York: Norton.

Fuchs, Stephan. 2001. Beyond agency. *Sociological Theory* 19 (1):24–40.

Fulsang, Deborah. 2002. Mom, I'm ready for school. *The Globe and Mail,* September 28, L1, L6.

Fuss, Diana. 1995. *Identification papers.* New York: Routledge.

Garrison, Ednie Kaeh. 2000. U.S. feminism—grrrl style! Youth (sub)cultures and the technologics of the third wave. *Feminist Studies* 26 (1):141–170.

Gee, James Paul. 2000–2001. Identity as an analytic lens for research in education. In *Review of research in education,* ed. W. Secada, 99–124. Washington, DC: American Educational Research Association.

George, Lianne. 2007. Why are we dressing our daughters like this? *Maclean's,* January 1, 37–45.

Gilligan, Carol. 1982. *In a different voice: Psychological theory and women's development.* Cambridge, MA: Harvard University Press.

Gilligan, Carol, Nona Lyons, and Trudy J. Hanmer. 1990. *Making connections: The relational worlds of adolescent girls at Emma Willard School.* Cambridge, MA: Harvard University Press.

Glazier, Lynn. 2004. *It's a girls' world.* National Film Board of Canada.

Gleeson, Kate, and Hannah Frith. 2004. Pretty in pink: Young women presenting mature sexual identities. In *All about the girl: Culture, power, and identity,* ed. A. Harris, 103–113. New York: Routledge.

Goffman, Erving. 1959. *The presentation of self in everyday life.* New York: Doubleday.

Gonick, Marnina. 2003. *Between femininities: Ambivalence, identity, and the education of girls.* Albany: State University of New York Press.

———. 2006. Between "girl power" and "reviving Ophelia": Constituting the neoliberal girl subject. *NWSA Journal* 18 (2):1–23.

Goodman, Barak. 2003. *The merchants of cool.* Frontline.

Gore, Jennifer. 1993. *The struggle for pedagogies: Critical and feminist discourses as regimes of truth.* New York: Routledge.

Greenfield, Lauren. 2002. *Girl culture.* San Francisco: Chronicle Books.

Griffin, Christine. 2004. Good girls, bad girls: Anglocentrism and diversity in the constitution of contemporary girlhood. In *All about the girl: Culture, power, and identity,* ed. A. Harris, 29–43. New York: Routledge.

Grosz, Elizabeth. 1994. *Volatile bodies: Toward a corporeal feminism.* Bloomington: Indiana University Press.

Hall, Stuart. 1977. Culture, the media and the "ideological effect." In *Mass communications and society,* ed. J. Curran, M. Gurevitch, and J. Woollacott, 315–348. London: Edward Arnold in association with the Open University Press.

———. 1978. *Policing the crisis: Mugging, the state, and law and order.* New York: Holmes & Meier.

———. 1981. Notes on deconstructing the "popular." In *People's history and socialist theory,* ed. R. Samuel, 227–239. London: Routledge & Kegan Paul.

———. 1990. Cultural identity and diaspora. In *Identity: Community, culture, difference,* ed. J. Rutherford, 222–237. London: Lawrence & Wishart.

———. 1996. Introduction: Who needs "identity"? In *Questions of cultural identity,* ed. S. Hall and P. Du Gay, 1–17. Thousand Oaks, CA: Sage.

Hall, Stuart, and Tony Jefferson. 1976. *Resistance through rituals: Youth subcultures in post-war Britain.* London: Hutchinson.

Haraway, Donna. 1988. Situated knowledges: The science question in feminism and privilege of partial perspective. *Feminist Studies* 14 (3):575–597.

Hardt, Michael, and Vittorio Negri. 2000. *Empire.* Cambridge, MA: Harvard University Press.

Harris, Anita. 2001. Revisiting bedroom culture: New spaces for young women's politics. *Hectate* 27 (1):128–139.

———, ed. 2004a. *All about the girl: Culture, power, and identity.* New York: Routledge.

———. 2004b. *Future girl: Young women in the twenty-first century.* New York: Routledge.

———. 2004c. Jamming girl culture: Young women and consumer citizenship. In *All about the girl: Culture, power, and identity,* ed. A. Harris, 163–172. New York: Routledge.

Havrilesky, Heather. 2002. Powerpuff girls meet world. *Salon.com,* http://www.salon.com/mwt/feature/2002/07/02/powerpuff/print.html.

Hays, Sharon. 1994. Structure and agency and the sticky problem of culture. *Sociological Theory* 12 (1):57–72.

Hebdige, Dick. 1979. *Subculture: The meaning of style.* London: Methuen.

———. 1988. *Hiding in the light: On images and things.* London: Routledge.

Hey, Valerie. 1997. *The company she keeps: An ethnography of girls' friendships.* Philadelphia, PA: Open University Press.

Heywood, Leslie, and Jennifer Drake, eds. 1997. *Third wave agenda: Being feminist, doing feminism.* Minneapolis: University of Minnesota Press.

Hodkinson, Paul. 2002. *Goth: Identity, style and subculture.* Oxford: Berg.

Holland, Dorothy C., and Margaret Eisenhart. 1990. *Educated in romance: Women, achievement, and college culture.* Chicago: University of Chicago Press.

Ige, Barbara Kaoru. 1998. For sale: A girl's life in the global economy. In *Millennium girls: Today's girls around the world,* ed. S. A. Inness, 45–60. Lanham, MD: Rowman & Littlefield Publishers.

Inness, Sherrie A., ed. 1998a. *Delinquents and debutantes: Twentieth-century American girls' cultures.* New York: New York University Press.

———, ed. 1998b. *Millennium girls: Today's girls around the world.* Lanham, MD: Rowman & Littlefield Publishers.

———, ed. 2000. *Running for their lives: Girls, cultural identity, and stories of survival.* Lanham, MD: Rowman & Littlefield Publishers.

Jefferson, Tony. 1975. Cultural responses of the Teds. In *Resistance through rituals: Youth subcultures in post-war Britain,* ed. S. Hall and T. Jefferson, 81–86. The Centre for Contemporary Cultural Studies, Birmingham: Cambridge University Press.

Jones, Alison. 1993. Becoming a "girl": Post-structuralist suggestions for educational research. *Gender & Education* 5 (2):157–166.

———. 1997. Teaching post-structuralist feminist theory in education: Student resistances. *Gender & Education* 9 (3):261–269.

Kamen, Paula. 2000. *Her way: Young women remake the sexual revolution.* New York: New York University Press.

Kearney, Mary Celeste. 1997. The missing links: Riot grrrl-feminism-lesbian culture. In *Sexing the groove: Popular music and gender*, ed. S. Whiteley. London, New York: Routledge.

———. 1998. Producing girls: Rethinking the study of female youth culture. In *Delinquents and debutantes: Twentieth-century American girls' cultures*, ed. S. A. Inness, 285–310. New York: New York University Press.

Kelly, Deirdre M. 1993. *Last chance high: How girls and boys drop in and out of alternative schools*. New Haven: Yale University Press.

———. 2000. *Pregnant with meaning: Teen mothers and the politics of inclusive schooling*. New York: Peter Lang.

Kelly, Deirdre M., and Shauna Pomerantz. 2007. Mean, wild, and alienated: Girls and the state of feminism in popular culture. Manuscript under review.

Kelly, Deirdre M., Shauna Pomerantz, and Dawn H. Currie. 2005. "You can't land an ollie properly in heels": Skater girlhood and emphasized femininity. *Gender & Education* 17 (3):229–249.

———. 2006. "No boundaries?" Girls' interactive, online learning about femininities. *Youth and Society* 38 (1):3–28.

Kenway, Jane, and Sue Willis with Jill Blackmore and Leonie Rennie. 1998. *Answering back: Girls, boys, and feminism in schools*. New York: Routledge.

Klein, Melissa. 1997. Duality and redefinition: Young feminism and the alternative music community. In *Third wave agenda: Being feminist, doing feminism*, ed. L. Heywood and J. Drake, 207–225. Minneapolis: University of Minnesota Press.

Labi, Nadya. 1998. Girl power. *Time.com*, http://www.time.com/time/magazine/1998/dom/980629/box1.html.

Lacan, Jacques. 1977. *Écrits*. New York: Norton.

Lather, Patti. 1991. *Getting smart: Feminist research and pedagogy with/in the postmodern*. New York: Routledge.

———. 2007. *Getting lost: Feminist efforts toward a double(d) science*. Albany: State University of New York Press.

Lather, Patti, and Chris Smithies. 1997. *Troubling the angels: Women living with HIV/AIDS*. Boulder, CO: Westview Press.

Laws, Cath, and Bronwyn Davies. 2000. Poststructuralist theory in practice: Working with "behaviourally disturbed" children. *International Journal of Qualitative Studies in Education* 13 (3):205–221.

Lees, Sue. 1986. *Losing out: Sexuality and adolescent girls*. London: Hutchinson.

Leonard, Marion. 1997. "Rebel girl, you are the queen of my world": Feminism, "subculture" and grrrl power. In *Sexing the groove: Popular music and gender*, ed. S. Whiteley, 230–255. London: Routledge.

Lesko, Nancy. 1988. The curriculum of the body: Lessons from a Catholic high school. In *Becoming feminine: The politics of popular culture*, ed. L. G. Roman, L. Christian-Smith, and E. Ellsworth, 123–142. London: Falmer Press.

Levy, Ariel. 2005. *Female chauvinist pigs: Women and the rise of raunch culture*. New York: Free Press.

Lewis, Lisa A. 1989. Consumer girl culture: How music video appeals to girls. In *Television and women's culture: The politics of the popular,* ed. M. E. Brown, 89–101. Sydney: Currency Press.

Luke, Carmen, ed. 1996. *Feminisms and pedagogies of everyday life.* Albany: State University of New York Press.

Luke, Carmen, and Jennifer Gore. 1992. *Feminisms and critical pedagogy.* New York: Routledge.

Lurie, Alison. 1981. *The language of clothes.* New York: Random House.

MacLeod, Jay. 1995. *Ain't no makin' it: Aspirations and attainment in a low-income neighborhood.* Boulder, CO: Westview Press.

Maglin, Nan Bauer, and Donna Marie Perry. 1996. *"Bad girls"/"good girls": Women, sex, and power in the nineties.* New Brunswick, NJ: Rutgers University Press.

Man, Guida. 2004. Gender, work and migration: Deskilling Chinese immigrant women in Canada. *Women's Studies International Forum* 27 (2): 135–149.

Mazzarella, Sharon R. 2005. *Girl wide web: Girls, the Internet, and the negotiation of identity.* New York: Peter Lang.

Mazzarella, Sharon R., and Norma O. Pecora, eds. 1999. *Growing up girls: Popular culture and the construction of identity.* New York: Peter Lang.

McNay, Lois. 1999. Gender, habitus and the field: Pierre Bourdieu and the limits of reflexivity. *Theory, Culture & Society* 16 (1):95–117.

———. 2000. *Gender and agency: Reconfiguring the subject in feminist and social theory.* Cambridge, UK: Polity Press.

McRobbie, Angela. 1991. *Feminism and youth culture: From "Jackie" to "Just seventeen."* Boston: Unwin Hyman.

———. 1996. Different, youthful, subjectivities. In *The post-colonial question: Common skies, divided horizons,* ed. I. Chambers and L. Curti. London: Routledge.

———. 1997. Second-hand dresses and the role of the ragmarket. In *The subcultures reader,* ed. K. Gelder and S. Thornton. London: New York: Routledge.

———. 2004. Notes on postfeminism and popular culture: Bridget Jones and the new gender regime. In *All about the girl: Culture, power, and identity,* ed. A. Harris. New York: Routledge.

McRobbie, Angela, and Jenny Garber. 1991. Girls and subcultures. In *Feminism and youth culture: From "Jackie" to "Just Seventeen,"* ed. Angela McRobbie, 1–15. Boston: Unwin Hyman.

Merten, Don E. 1997. The meaning of meanness: Popularity, competition, and conflict among junior high school girls. *Sociology of Education* 70 (July):175–191.

Milner, Murray. 2004. *Freaks, geeks, and cool kids: American teenagers and the culture of consumption.* New York: Routledge.

Mitchell, Claudia, and Jacqueline Reid-Walsh, eds. 2005. *Seven going on seventeen: Tween studies in the culture of girlhood.* New York: Peter Lang.

Mitchell, Juliet. 1984. *Women, the longest revolution: Essays on feminism, literature and psychoanalysis.* London: Virago.

Morris, Meaghan. 1993. Things to do with shopping centres. In *The Cultural Studies Reader,* ed. S. During, 296–319. London: Routledge.

Mulvey, Laura. 1975. Visual pleasure and narrative cinema. *Screen* 16 (3):6–18.

Orenstein, Peggy. 1994. *Schoolgirls: Young women, self-esteem, and the confidence gap.* New York: Doubleday.

Ortner, Sherry. 1998. Identities: The hidden life of class. *Journal of Anthropological Research* 54 (1):1–17.

Owen, Susan A. 1999. Vampires, postmodernity, and postfeminism: Buffy the vampire slayer. *Journal of Popular Film and Television* 27 (2):24–28.

Pipher, Mary Bray. 1994. *Reviving Ophelia: Saving the selves of adolescent girls.* New York: Putnam.

Pomerantz, Shauna. 2006. "Did you see what she was wearing?" The power and politics of schoolgirl style. In *Girlhood: Redefining the limits,* ed. Y. Jiwani, C. Steenbergen, and C. Mitchell, 173–190. Montreal, Canada: Black Rose Books.

———. 2007. Cleavage in a tank top: Bodily prohibition and the discourses of school dress codes. *Alberta Journal of Educational Review* 53 (4): 373–386.

Pomerantz, Shauna, Dawn H. Currie, and Deirdre M. Kelly. 2004. Sk8er girls: Skateboarders, girlhood, and feminism in motion. *Women's Studies International Forum* 27 (5–6):547–557.

Pyke, Karen, and Tran Dang. 2003. "FOB" and "whitewashed": Identity and internalized racism among second generation Asian Americans. *Qualitative Sociology* 26 (2):147–171.

Quindlen, Anna. 1996. And now, babe feminism. In *"Bad girls"/"good girls": Women, sex, & power in the nineties,* ed. N. B. Maglin and D. Perry, 3–5. New Brunswick, NJ: Rutgers University Press.

Radway, Janice A. 1984. *Reading the romance: Women, patriarchy, and popular literature.* Chapel Hill, NC: University of North Carolina Press.

Reid-Walsh, Jacqueline, and Claudia Mitchell. 2004. Girls' web sites: A virtual "room of one's own"? In *All about the girl: Culture, power, and identity,* ed. A. Harris, 173–182. New York: Routledge.

Richardson, Laurel. 2003. Writing: A method of inquiry. In *Collecting and interpreting qualitative materials,* 2nd edition, ed. N. K. Denzin and Y. S. Lincoln, 499–541. Thousand Oaks, CA: Sage.

Riordan, Ellen. 2001. Commodified agents and empowered girls: Consuming and producing feminism. *Journal of Communication Inquiry* 25 (3): 279–297.

Roman, Leslie G., and Michael W. Apple. 1990. Is naturalism a move away from positivism? Materialist and feminist approaches to subjectivity in ethnographic research. In *Qualitative inquiry in education,* ed. E. Eisner and Peshkin, 38–73. New York: Teacher's College Press.

Ruditis, Paul. 2005. *Rainbow Party.* New York: Simon Pulse.

Said, Edward W. 1978. *Orientalism*. New York: Pantheon Books.

Schilt, Kristen. 2003. "I'll resist with every inch of my breath": Girls and zine making as a form of resistance. *Youth & Society* 35 (1):71–97.

Schrum, Kelly. 2004. *Some wore bobby sox: The emergence of teenage girls' culture, 1920–1945*. New York: Palgrave Macmillan.

Silverman, Kaja. 1994. Fragments of a fashionable discourse. In *On fashion*, ed. S. Benstock and S. Ferriss, 183–196. New Brunswick, NJ: Rutgers University Press.

Simmons, Rachel. 2002. *Odd girl out: The hidden culture of aggression in girls*. New York: Harcourt.

Skeggs, Beverley. 1997. *Formations of class and gender: Becoming respectable*. London: Sage.

Smith, Dorothy. 1987. *The everyday world as problematic: A feminist sociology*. Toronto: University of Toronto Press.

———. 1988. Femininity as discourse. In *Becoming feminine: The politics of popular culture*, ed. L. Roman, L. Christian-Smith, and E. Ellsworth, 37–59. London: Falmer Press.

Smith, Linda Tuhiwai. 1999. *Decolonizing methodologies: Research and indigenous peoples*. Dunedin, New Zealand: University of Otago Press.

Søndergaard, Dorte Marie. 2002. Poststructuralist approaches to empirical analysis. *Qualitative Studies in Education* 15 (2):187–204.

Spivak, Gayatri Chakravorty. 1974. Translator's preface. In *Of Grammatology*, ed. Jacques Derrida, xi-lxxxvii. Baltimore, MD: Johns Hopkins University Press.

St. Pierre, Elizabeth Adams. 2000. Poststructural feminism in education: An overview. *International Journal of Qualitative Studies in Education* 13 (5): 477–515.

St. Pierre, Elizabeth Adams, and Wanda S. Pillow. 2000. *Working the ruins: Feminist poststructural theory and methods in education*. New York: Routledge.

Stacey, Judith. 1990. Sexism by a subtler name? Poststructural conditions and postfeminist consciousness in Silicon Valley. In *Women, class, and the feminist imagination: A socialist feminist reader*, ed. K. V. Hansen and I. J. Philipson, 338–356. Philadelphia: Temple University Press.

Statistics Canada. 2001. Vancouver Community Profile, http://www.statcan .ca/start.html.

Stepp, Laura Sessions. 2002. Nothing to wear. *The Washington Post*, June 3, C01.

Storey, John. 2006. *Cultural theory and popular culture: A reader*. 3rd ed. Toronto: Pearson/Prentice Hall.

Taft, Jessica K. 2004. Girl power politics: Pop-culture barriers and organizational resistance. In *All about the girl: Culture, power, and identity*, ed. A. Harris, 69–78. New York: Routledge.

Thornton, Sarah. 1994. Moral panic, the media and British rave culture. In *Microphone fiends*, ed. A. Ross and T. Rose, 176–192. New York: Routledge.

————. 1997. The social logic of subcultural capital. In *The subcultures reader,* ed. K. Gelder and S. Thornton, 200–212. London: Routledge.

Tolman, Deborah L., and Tracy E. Higgins. 1994. How being a good girl can be bad for girls. In *"Bad girls"/"good girls": Women, sex, and power in the nineties,* ed. N. B. Maglin and D. Perry, 205–225. New Brunswick, NJ: Rutgers University Press.

Trebay, Guy. 2003. The skin wars start earlier and earlier. *NYTimes.com,* http://query.nytimes.com/gst/fullpage.html?res=9C01EEDF1538F931 A3575AC0A9659C8B63&partner=rssnyt&emc=rss.

Trinh, T. Minh-Ha. 1989. *Woman, native, other: Writing postcoloniality and feminism.* Bloomington: Indiana University Press.

Valverde, Mariana. 1991. As if subjects existed: Analysing social discourses. *Canadian Review of Sociology and Anthropology* 28 (2):173–187.

Visweswaran, Kamala. 1994. *Fictions of feminist ethnography.* Minneapolis: University of Minnesota Press.

Wald, Gayle. 1998. Just a girl? Rock music, feminism, and the cultural construction of female youth. *Signs: Journal of Women in Culture and Society* 23 (3):585–610.

Walker, Rebecca, ed. 1995. *To be real: Telling the truth and changing the face of feminism.* New York: Doubleday.

Walkerdine, Valerie. 1990. *Schoolgirl fictions.* London: Verso.

————. 1993. Girlhood through the looking glass. In *Girls, girlhood and girls' studies in transition,* ed. M. de Ras and M. Lunenberg, 9–24. Amsterdam: Het Spinhuis.

Walkerdine, Valerie, Helen Lucey, and June Melody. 2001. *Growing up girl: Psychosocial explorations of gender and class.* Houndmills, UK: Palgrave.

Warwick, Alexandra, and Dani Cavallaro. 1998. *Fashioning the frame: Boundaries, dress and body.* London: Berg.

Weber, Sandra, and Claudia Mitchell. 1995. *That's funny, you don't look like a teacher! Interrogating images and identity in popular culture.* London: Falmer Press.

Weedon, Chris. 1987. *Feminist practice and poststructuralist theory.* Oxford: B. Blackwell.

West, Cornel. 1995. A matter of life and death. In *The identity question,* ed. J. Rajchman, 15–19. New York: Routledge.

Wilkins, Amy C. 2004. "So full of myself as a chick": Goth women, sexual independence, and gender egalitarianism. *Gender & Society* 18 (3):328–349.

Willis, Paul E. 1977. *Learning to labor: How working class kids get working class jobs.* Morningside Books. New York: Columbia University Press.

Wilson, Elizabeth. 1985. *Adorned in dreams: Fashion and modernity.* London: Virago.

Winship, Janice. 1987. *Inside women's magazines.* London: Pandora.

Wiseman, Rosalind. 2002. *Queen bees & wannabes: Helping your daughter survive cliques, gossip, boyfriends, and other realities of adolescence.* New York: Crown Publishers.

Woodend, Dorothy. 2002. Girl power. *The Vancouver Sun,* November 23, H8, H9.

Yon, Daniel A. 2000. *Elusive culture: Schooling, race, and identity in global times.* Albany: State University of New York Press.

Young, Iris Marion. 1990. Women recovering our clothes. In *Throwing like a girl and other essays in feminist philosophy and social theory.* Bloomington, IN: Indiana University Press.

INDEX